SpringerBriefs in Electrical and Computer Engineering

For further volumes:
http://www.springer.com/series/10059

Yongkang Liu • Xuemin (Sherman) Shen

Cognitive Resource Management for Heterogeneous Cellular Networks

 Springer

Yongkang Liu
University of Waterloo
Waterloo, ON, Canada

Xuemin (Sherman) Shen
Department of Electrical
  and Computer Engineering
University of Waterloo
Waterloo, ON, Canada

ISSN 2191-8112   ISSN 2191-8120 (electronic)
ISBN 978-3-319-06283-9   ISBN 978-3-319-06284-6 (eBook)
DOI 10.1007/978-3-319-06284-6
Springer Cham Heidelberg New York Dordrecht London

Library of Congress Control Number: 2014937897

Printed on acid-free paper

Springer is part of Springer Science+Business Media (www.springer.com)

Preface

Smartphone fever along with roaring mobile traffic poses great challenges for today's cellular networks. In general, cellular operators and vendors promise to provide seamless mobile access to end users. However, given the temporal and spatial variations of the ever increasing smartphone user demand, they feel obligated to deploy cellular access nodes in a more flexible and intelligent way, e.g., plug-to-play, self-organized, and cost-effective. Therefore, various types of small cells are being adopted indoors and outdoors to complement macrocells in cellular hotspots and blind zones. Designed by different purposes of operations, macrocells and small cells show heterogeneous characteristics, which requires new techniques to solve the challenging coexistence issues including interference management, cell coordination, and interworking.

In this Brief, we exploit cognitive radio techniques to improve spectrum utilization and perform flexible network management in the heterogeneous cellular network (HetNet) formed by macrocells and small cells. Background and literature survey of HetNet and cognitive radio techniques are first presented in Chap. 1. We then introduce an open cell management framework in Chap. 2, namely as *cognitive cellular network management* (CCN), which is mainly aimed to improve spectrum utilization and mitigate co-channel interference in HetNet. In Chap. 3, we investigate in wireless backhaul for flexible deployment of small cells, which requires smooth and reliable communications with the network controller even if wired portal is not available. Instead of static spectrum allocation, overlay spectrum reuse fits such need better, which accommodates the backhaul traffic by fully utilizing the intermittent spare spectrum resources with small spatial prints. An opportunistic routing protocol for wireless backhaul is presented along with the introduction of joint channel and relay selection. In Chap. 4, we further address on the coexistence issue between macrocells and small cells. When small cells are loosely controlled due to limited bandwidth in dense deployment, the effective allocation of radio resources becomes challenging. We propose a distributed QoS-aware cognitive MAC scheme which facilitates users in small cells to find available resources in an opportunistic way so that they can transmit at higher power for better link quality while maintaining tolerable interference to macrocell transmissions.

In addition, a penalty approach is used in backhaul to secure the effectiveness of power allocation in the transmission channels. Finally, we summarize the Brief and provide future research directions in Chap. 5.

We would like to thank our BBCR colleagues at the University of Waterloo for their valuable comments and suggestions on the Brief. We would also like to thank the Springer Editors Susan Lagerstrom-Fife and Jennifer Malat for their great help in getting this Brief published. This research work was financially supported by NSERC, Canada.

Waterloo, ON, Canada   Yongkang Liu
Waterloo, ON, Canada   Xuemin (Sherman) Shen

Contents

1 **Introduction** 1
   1.1 Heterogeneous Cellular Networks 2
      1.1.1 Small Cell Deployment... 3
      1.1.2 Vertical Cellular Access Architecture........................... 5
      1.1.3 Challenges.. 6
      1.1.4 Related Work... 8
   1.2 Cognitive Radio Networks 9
   References 11

2 **Cognitive Cellular Network Management** 13
   2.1 CCN Framework 13
   2.2 Applications and Challenges 14
      2.2.1 Femtocell Deployment 14
      2.2.2 Resource Management in HetNet............................... 15
      2.2.3 Backhaul Bottleneck Mitigation 16
   2.3 Research Topics 17
      2.3.1 Wireless Backhaul Routing 17
      2.3.2 Interference Management 19
   2.4 Summary 21
   References 22

3 **Spectrum Aware Opportunistic Routing for Wireless Backhaul** 25
   3.1 System Model 26
   3.2 Spectrum Aware Opportunistic Routing 27
      3.2.1 Protocol Overview 27
      3.2.2 Routing Protocol Analysis 29
   3.3 Joint Channel and Relay Selection 34
      3.3.1 Novel Routing Metric 34
      3.3.2 Heuristic Algorithm 35
   3.4 Simulation Results 38
      3.4.1 Simulation Settings 39
      3.4.2 PU Activities 40

        3.4.3   Multi-User Diversity... 42
        3.4.4   Effectiveness of Routing Metric 43
    3.5   Summary 44
    References 45

4   **QoS-Aware Cognitive MAC and Interference Management**
    for HetNet.. 47
    4.1   System Model 48
        4.1.1   Network Model 48
        4.1.2   Traffic Model... 48
    4.2   QoS-Aware Cognitive MAC for Small Cells 49
        4.2.1   Channel Sensing 50
        4.2.2   Service Differentiation 52
        4.2.3   Performance Analysis 52
    4.3   Power Allocation Under Violation Penalty 54
        4.3.1   Effective Control in Constrained Backhaul..................... 54
        4.3.2   Game Theoretic Power Allocation.............................. 56
    4.4   Simulation Results 57
        4.4.1   Simulation Settings 57
        4.4.2   Delay of Homogeneous Traffic 57
        4.4.3   Delay of Heterogeneous Traffic 58
        4.4.4   Performance of Service Differentiation....................... 59
        4.4.5   Power Allocation Under Violation Penalty 60
    4.5   Summary 60
    Appendix: Proof of Nash Equilibrium in Sect. 4.3 61
    References 62

5   **Conclusions and Future Directions** 63
    5.1   Conclusions 63
    5.2   Future Research Directions... 64
    References 65

Acronyms

ASP	Arbitrary sensing period
BS	Base station
CBR	Constant bit rate
CCC	Common control channel
CCN	Cognitive cellular network management
CDF	Cumulative distribution function
CoMP	Coordinated multipoint
CRN	Cognitive radio network
CSMA/CA	Carrier sense multiple access with collision avoidance
CTT	Cognitive transport throughput
DSL	Digital subscriber line
FBS	Femtocell base station
FCC	Federal Communications Commission
HDV	High definition video
HetNet	Heterogeneous cellular networks
IEEE	Institute of Electrical and Electronics Engineers
ISM	Industrial, scientific and medical
ITU	International Telecommunication Union
LTE	Long term evolution
LTE-A	LTE advanced
MAC	Media access control
MBS	Macrocell base station
MIMO	Multiple-input multiple-output
NE	Nash equilibrium
OCR	Opportunistic cognitive routing
OFDMA	Orthogonal frequency-division multiple access
PDF	Probability density function
POMDP	Partially observable Markov decision process
POP	Point of presence
PU	Primary user
QoS	Quality of service

RAN	Radio access networks
RF	Radio frequency
RREQ	Routing request
RRSP	Routing response
SCF	Small cell forum
SBS	Small cell base station
SINR	Signal to noise and interference ratio
SNSINV	Sensing invitation
SU	Secondary user
UWB	Ultra-wideband
VBR	Variable bit rate
WiMAX	Worldwide interoperability for microwave access
3GPP	3rd generation partnership program

Chapter 1
Introduction

The fast development of cellular communications corroborates the success of the mobile Internet which penetrates into our daily lives by connecting end user devices to the Internet with diverse services of qualities. Thanks to the powerful computation and communication hardware platform on the mobile devices, cellular users can subscribe more data hungry services and exchange more data than ever before. For example, a smartphone user generates on average as much as 49 times data of that of a voice-only cellphone. A fourth-generation (4G) connection generated 14.5 times more traffic on average than a non-4G connection in 2013. It is estimated that the global mobile devices generated 18 extrabytes of data traffic in 2013, i.e., nearly 18 times the size of the entire global Internet in 2000 [1]. It is critical to improve the network capacity and meet the explosively growing data demand.

In cellular networks, two approaches dominate in the evolutions of techniques to meet the ever-increasing bandwidth demands of cellular users: "expanding" and "digging deeper". The first approach straightforwardly investigate more in cellular infrastructure and obtains more spectrum resources at the expense of billions of dollars. For example, 4G cellular networks are typically placing more macrocell sites in the same serving area than 2G and 3G cellular networks, which is referred to as "network densification" techniques, as well as more frequency bands are assigned for 4G usage. However, since appropriate cites for tower-like base stations (BSs) and available radio bands are inherently limited and very expensive, such an additive construction can not catch up with the exponential growth of user demand. Moreover, inter-cell interference becomes more notable in the denser deployment cases, which prohibits the size of macrocell from further shrinking and compromises the marginal gain due to frequency reuse and interference management cost. Therefore, the operators usually turn to the second approach, i.e., deploying new physical and link layer techniques to improve spectrum utilization in the existing bands and network infrastructure. Advanced techniques, such as multiple-input multiple-output (MIMO), high order modulation and smart antennas, have been applied in the cells working with scheduling schemes designed for multi-dimensional resource allocation, such as in orthogonal frequency-division multiple access (OFDMA)

Y. Liu and X. Shen, *Cognitive Resource Management for Heterogeneous Cellular Networks*, SpringerBriefs in Electrical and Computer Engineering, DOI 10.1007/978-3-319-06284-6_1, © The Author(s) 2014

systems, to effectively improve the network performance. However, continued enhancement is not always the cost-efficient choice as these techniques usually have high operational complexity and maintenance cost. Aforementioned evolutional efforts cannot fully solve the bandwidth shortage. To this end, a simple yet efficient solution is required to improve spectrum utilization and dimension the future system design and management.

1.1   Heterogeneous Cellular Networks

In cellular radio access networks (RAN), a mesh of macrocells is initially deployed to cover the serving area to provide public cellular access for subscribers as shown in Fig. 1.1a. To meet the growing user demand, increasingly densifying macrocell to enhance the network capacity does not always work because of fewer cell site options, more frequent handoff of mobile users and increasing co-channel interference. Meanwhile, the user demands are not distributed evenly in the network, nor is the macrocell signal strength at the receiver. An alternative solution by applying tiered network infrastructure in RAN proves to be feasible and beneficial [2]. As shown in Fig. 1.1b, built upon existing macrocells, low power low cost cellular access nodes are deployed to serve cellular users at the spots where macrocell base stations (MBSs) can not provide effective or efficient coverage. As there are different types of low power access nodes to meet various deployment requirements, cellular networks are mitigating into a heterogenous access infrastructure, commonly known as heterogeneous cellular networks (HetNet) [3].

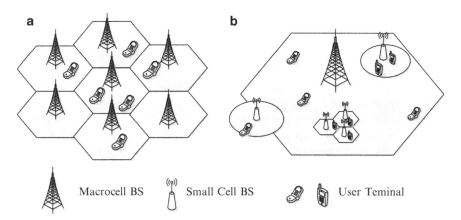

Fig. 1.1 RAN infrastructures in cellular networks: (**a**) flat (**b**) tiered

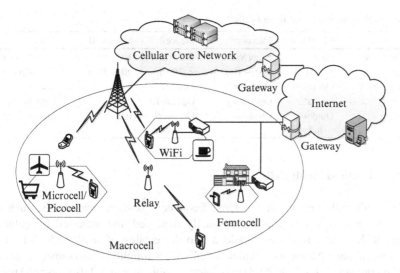

Fig. 1.2 Heterogeneous cellular networks

1.1.1 Small Cell Deployment

In HetNet, a low power access node usually serves a smaller area and has simpler structure and functions in the device compared with a MBS. The low power access node and the users in the serving area form a small cell coined by the relatively smaller communication range, in which the access node is referred to as small cell base station (SBS). Small cells are deployed by cellular operators or end users at the locations within the macrocell to provide public or private access for particular purposes, e.g., to improve network capacity in hotspots, to compensate the long distance loss for users at cell edge or to extend coverage to the blind zone, as shown in Fig. 1.2. According to the working frequency and the deployment and control schemes, small cells can be roughly categorized into two types:

1.1.1.1 Out-of-Band Small Cells

These small cells operate in the frequency bands other than the licensed frequency bands of macrocells, e.g., the unlicensed 2.4 GHz industrial, scientific and medical (ISM) band. Out-of-band small cells are usually deployed by end users, e.g., WiFi hotspots. Nowadays, cellular operators become more interested in deploying enterprise-grade WiFi networks to offload cellular data from MBS to WiFi access points. The operator-deployed WiFi networks are usually open to their own subscribers only.

Table 1.1 Specifications of different cells

	Macrocell	Microcell	Picocell	Femtocell	WiFi
Transmit power	50 W	a few Watts	>200 mW	$10 \sim 100$ mW	$100 \sim 200$ mW
Range	$1 \sim 5$ km	$300 \sim 1000$ m	< 200 m	$20 \sim 30$ m	$100 \sim 200$ m
Deployment	Operator	Operator	Operator	User	User
Operating bands	Operator's	Operator's	Operator's	Operator's	Unlicensed
Coverage	Outdoor/ indoor	Outdoor	Outdoor/ indoor	Indoor	Indoor

1.1.1.2 In-Band Small Cells

These small cells operate in the same frequency bands as macrocells, such as microcells and picocells which are usually deployed and managed by cellular operators. Recently, femtocell is added into the catalog of small cells, which is aimed to enhance the indoor cellular signal with a simplified BS connected to the cellular core network via the third party Internet cable service [2]. Because end users can deploy their own femtocells in their houses, cellular operators only have limited control over these private access femtocells, which makes co-channel interference management and cell coordination more challenging.

A brief summary of the specifications of small cells is listed in Table 1.1. Typical applications of SBS include micro/pico cells deployed outdoors in urban streets, the indoor cellular signal penetration at airports or shopping malls, private home femtocells and WiFi access points that offload data traffic from MBS [4]. Each SBS obtains a unique cell identification (ID) as MBS from the operator to provision users to access.[1] Users select the serving BS according to the measurement of the real time link quality.

Compared with macrocells, small cells have unique advantages in some usage scenarios, such as capacity enhancement in hotspots, and coverage expansion into the houses and workplaces. Basically, small cells allow for flexible SBS deployment and simple transceiver design due to the limited communication coverage. As the radio environment is becoming more and more complicated, using small cells is beneficial for the operators to deal with the localized coverage and link enhancement while offloading the traffic from the macrocells to femtocells or WiFi [1].

---

[1]The operator may deploy distributed antenna units, e.g., remote radio units (RRU) in 3GPP [5], or relay nodes to expand the area a single MBS covers or enhance the signal performance at the spot of interest within a macrocell. However, these access devices are simply physical relay amplifiers, and they are still affiliated with the MBS and using the cell ID of the macrocell where they reside in. The connections in the macrocell are centrally controlled by MBS, the transitions of which are transparent to users.

1.1.2 Vertical Cellular Access Architecture

In HetNet, the introduction of small cells promotes vertical cellular access architecture to be diverse, which provides more options for end users to access the cellular infrastructure. As shown in Fig. 1.3, RAN and the core network exchange data and control signaling messages via the gateway on the edge. End users in RAN are connected to the central controller located in the core network by a route consisting of two types of links, which are radio access links and RAN backhaul links, respectively. Specifically, radio access links are the direct wireless links formed in the frequency bands between users and access nodes in macrocells and small cells. While RAN backhaul, or *backhaul*, represents the links connecting access nodes to the gateway of core network, which acts as a bridge of aggregating (distributing) upstream (downstream) traffic/signaling from (to) individual cells [6, 7]. Backhaul links can either reuse the fibre points of presence (POPs) of the existing macrocells or "self-backhaul" over radio links [7, 8].

Vertical access in a tiered infrastructure enables the operators and end users with more flexibility to decide when and where to deploy small cells if MBS can not meet the service requirement in a cost-effective way. Traditional RAN with flat deployment of macrocells only allow one type of radio access links, i.e., the ones between users and MBS. MBSs are usually installed on the preplanned sites with high speed cable network as backhaul. Once a user attaches an MBS, the route between user and central controller is determined. Meanwhile, the cellular connection of the user is

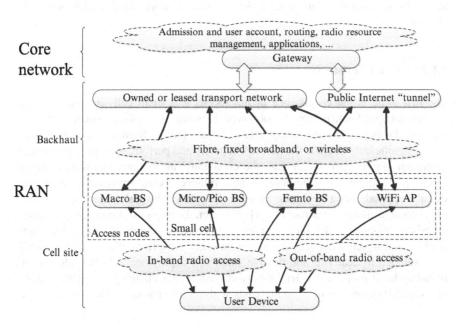

Fig. 1.3 Vertical cellular architecture in HetNet

also "deterministic" in the macrocell according to the settings on available services and performance made by cellular operators. For example, on a congested urban site, the cellular operator has to set strict download/upload speed limits and disable heavy data usage so that MBS can serve more users with mild traffic demand, which may impair the experience of some mobile services [9]. On the contrary, cellular operators can correspondingly place different small cells in HetNet to provide a comprehensive access solution in a region based upon the statistics of ambient radio environment and user demand. On the other hand, end users in HetNet have more choices on the route, not only in radio access links but also in backhaul links, i.g., how to determine the way that access nodes connect with the gateway. In terms of the combination of different types of radio access links and backhaul links, users can better match their needs and budgets with the advertised services from different options. Take the example of the data hunger mentioned above, cost sensitive users can turn to subscribe public WiFi access [10] while other users can switch to local femtocells in their houses or offices to keep enjoying high speed cellular services with higher bandwidth and volume cap [4].

1.1.3 Challenges

Recognizing the differences between macrocells and small cells, the implementation of small cells along with existing macrocells encounters new challenges. It is necessary to re-visit resource management and network management issues in HetNet.

1.1.3.1 Cell Coexistence

The first issue is the coexistence of macrocells and small cells in RAN. Severe co-channel interference may be observed between the in-band small cells and the neighboring cells, i.e., macrocells and the other in-band small cells, which would significantly degrade the network capacity and performance. In conventional macrocell deployment, to mitigate the co-channel interference between adjacent macrocells and optimize the network performance, frequency reuse schemes are usually applied along with the optimal BS placement calculated by network planning algorithms in macrocells [11]. However, the impact of small cell operations on existing macrocells and the entire network performance depends on the particular deployment, which is not yet clear in current study. Adaptive online schemes should be developed to deal with the flexible small cell deployment. On the other hand, in out-of-band small cells, HetNet may still encounter challenges in load balance although SBSs operate in separate bands without interference to/from macrocells.

In such a case, as cellular operators cannot rely on a fully centralized control to schedule traffic at both MBS and WiFi AP in a real time manner, distributed load balance coordination remains an open research issue.

1.1.3.2 Network Management

Small cells and vertical cellular access would complicate the network planning and control process in HetNet. Generally, the challenges are of threefold: (1) deviation of local utility from entire network performance, (2) loose control over individual cells, and (3) "leverage effects" in dense deployment of small cells.

Utility Deviation

In HetNet, cells in different layers may serve different types of users for various services, especially when private access small cells are only open to a private user group while competing with macrocells in the vicinity for bandwidth. As "rational players", these small cells intend to operate to maximize their local utility, which may notably deviate from the entire network performance. In addition, some small cells are deployed as an alternative in the fast varying ambient radio environment in which the observation at the central controller may not catch the instant local link quality to update the utility function.

Loose Control

The diversity of the backhaul networks indicates that the operators can no longer uniformly take full control of small cells in a tiered RAN infrastructure like they succeeded in macrocells. In some cases, the backhaul is even built out of the operator's infrastructure. For example, femtocell base stations (FBSs) may detour via the Internet to reach the cellular core network through the leased infrastructure from third party internet service providers (ISPs), which encumbers the real time centralized control.

Leverage Effects

In conventional cellular networks, macrocells can handle the occasional fluctuation of traffic or channel quality in the serving area, e.g., the short term demand increase due to handoff of mobile users from the neighboring macrocells. In HetNet, small cells would be densely deployed to serve a notable fraction of the user population, typically in urban area. When failure occurs in a single small cell, MBS and backhaul would experience more notable fluctuation in both volume and frequency, which is mainly caused by the accumulation of the jitters at individual small cells.

It greatly challenges the robustness of cellular networks when MBS is reluctant to accommodate so many handoff requests from small cells with the limited reserved bandwidth.[2]

1.1.3.3 Smart User

In today's ecosystem of cellular networks, which are comprised of both cellular operators and users, end users are more actively participating into networking and resource allocation to secure their perceived quality of service (QoS). For example, to improve the communication quality in an indoor environment, end users could deploy femtocells which operate in the licensed spectrum [2]. Therefore, the performance of HetNet becomes more dependent on the end users' operations, which requires end users should share more observations (e.g., link qualities, QoS, ambient radio environment and peer node contention level) with the central controller to make adaptive control and scheduling. Besides, different types of cellular cells have different capabilities of serving users. Although the cell selection can achieve diversity gain from a set of multiple visible cells, a large set usually means heavy cost in network coordination, e.g., increased airtime, more energy consumption and longer delay [13]. Users are expected to make "smarter" cell selection, i.e., the proper selection of the serving cells according to the characteristics of user demand (e.g., bandwidth, delay and budget) and cognition of circumstances (e.g., available cells in terms of type and cell site). To initiate the coordinated multipoint (CoMP) reception from multiple access nodes, users should calibrate the downlink links [14]. In addition, with the launch of new mobile services, such as e-health and personal financial services, critical QoS, security and privacy issues are also arising with higher requirements on end users [15].

1.1.4   Related Work

Landström et al. have first coined the concept of HetNet and explained that HetNet can increase the cellular network capacity by adding low power network access nodes [3]. Currently, the integration of small cells in the infrastructure have been concerned in the latest 3G/4G cellular systems including long term evolution (LTE)/LTE-Advanced (LTE-A) [14, 16, 17] of 3rd generation partnership program (3GPP) and worldwide interoperability for microwave access (WiMAX) [18], respectively. Andrews in [19] explains seven major ways that small-cell oriented HetNet are different than traditional tower-based cellular networks. Dhillon et al.

---

[2]In some cases, the catastrophic failures in the small cell layer may even gridlock the entire network, e.g., when hundreds of users turn to access MBS if SBSs suffer from software failure or power grid failure, analog to the congestion caused by the saturate attack [12].

in [13] illustrate a K-tier network structure of HetNet and discussed on the associated research challenges in the network modeling and capacity analysis. Yun and Shin in [20] propose an adaptive interference management framework of OFDMA femtocells for co-channel deployment with macrocells. They loop resource management and power control in a way as the "onion rings" with respective time scales and control resolution. Furthermore, there are also active discussions on the other technical and research issues, such as interworking and mobility management, handoff, network selection and load balancing in HetNet. Andrews, Conti and Shen have compiled a list of best readings in multi-tier cellular on behalf of the IEEE Communications Society as a comprehensive survey on the state-to-art techniques and standardization efforts in HetNet [21].

1.2 Cognitive Radio Networks

Cognitive radio networks (CRN) can provision the study of implementing HetNet because the researchers have developed many promising solutions for the similar challenges raised from the inherent flexible spectrum and network structure with the coexistence issues of different network nodes. In CRN, wireless nodes are divided into two separate user groups with different access priorities, i.e., primary users (PUs) and secondary users (SUs), respectively. Each user group forms a subnetwork layer in CRN with the unique services and the corresponding utility function in the frequency bands. Prioritized channel access is adopted to coordinate the transmissions of different user groups. As the licensed users, PUs have the exclusive priority in channel access. On the contrary, SUs are the unlicensed users. To protect the licensed transmissions of PUs, SUs need to opportunistically access the time-frequency resource blocks where PUs are inactive and operate with strict constraints to protect the active PUs' QoS, e.g., the maximum interference allowed at the PU's receiver. Figure 1.4 illustrates one example.

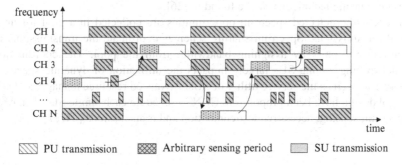

Fig. 1.4 The illustration of prioritized multi-channel access architecture in CRN

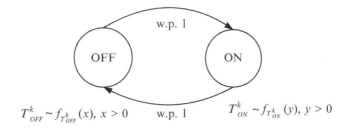

Fig. 1.5 Channel usage pattern

The free channels usually span over the spectrum and open a limited time window for unlicensed transmissions, which are referred to as spectrum access opportunities. The terms, e.g., spectrum holes and "white space" coined by the FCC [22], are also used exchangeably to denote the temporally available channels. Generally, from an SU's point of view, the channel availability for unlicensed transmissions can be modeled by an alternating ON/OFF process. The statistics of the states characterize the usage model of a channel, denoted as channel usage pattern, which is shown in Fig. 1.5. As a general definition [23], the sojourn time of an OFF period, when the channel k is free of PUs and available for secondary channel access, can be modeled as a random variable T_{OFF}^k with the probability density function (PDF) $f_{T_{OFF}^k}(x)$, $x > 0$. Similarly, the PDF of the duration of an ON period, when the channel is busy, is given in the term of $f_{T_{ON}^k}(y)$, $y > 0$.

CRN utilizes spectrum sensing to detect and track spectrum access opportunities for SUs. Sensing technique varies with signal processing schemes, such as energy detection, matched filtering, and feature detection [24]. Energy detection with the threshold-based decision is widely used because of its low computational and implementation complexities [25]. With energy detection, SUs sense the presence/absence of PUs based on the energy level of the received signals without any priori knowledge of PUs' signals. A detailed introduction on the existing spectrum sensing techniques can be found in [26].

In CRN, the detected spectrum opportunities are collected in a pool for unlicensed transmissions. Spectrum sharing schemes coordinate the transmissions of concurrent SU pairs by assigning bandwidth from the pool. The coordination and networking among SUs largely impacts the utilization of dynamic spectrum resources as well as the service utility of SUs. Advanced resource management and protocol design have been proposed in CRN to schedule each single SU's behavior in spectrum sensing, access request/contention and transmissions [24, 27].

References

1. Cisco (2014) Global mobile data traffic forecast update, 2013–2018. In: Cisco visual networking index. Cisco, San Jose. http://www.cisco.com/c/en/us/solutions/collateral/service-provider/visual-networking-index-vni/. Accessed 20 Feb 2014
2. Andrews JG, Claussen, H, Dohler M, Rangan S, Reed MC (2012) Femtocells: past, present, and future. IEEE J Sel Area Commun 30(3):497–508
3. Landström S, Furuskär A, Johansson K, Falconetti L, Kronestedt F (2011) Heterogeneous networks: increasing cellular capacity. In: Ericsson review. Ericsson. http://www.ericsson.com/res/thecompany/docs/publications/ericsson_review/2011/heterogeneous_networks.pdf. Accessed 27 Feb 2014
4. Small Cell Forum (2014) Document 103.03.01 overview. In: Release three: urban foundations. Small Cell Forum. http://www.scf.io/en/documents/103_-_Urban_Small_Cells_Release_overview.php. Accessed 27 Feb 2014
5. Yu Y, Luo Y, Gu D (2010) MU-MIMO downlink transmission strategy based on the distributed antennas for 3GPP LTE-A. In: Proceedings of IEEE Globecom' 10, Miami
6. Raza H (2011) A brief survey of radio access network backhaul evolution: part I. IEEE Comm Mag 49(6):164–171
7. Raza H (2013) A brief survey of radio access network backhaul evolution: part II. IEEE Comm Mag 51(5):170–177
8. Donegan P (2012) Small cell backhaul: what, why and how? In: Tellabs white paper. Heavy Reading. http://www.tellabs.com/resources/papers/tlab_smallcellbackhaul_wh.pdf. Accessed 27 Feb 2014
9. Pepitone J (2012) AT&T raises limit for smartphone data slowdown. In: CNN money. CNN. http://money.cnn.com/2012/03/01/technology/att_data_slowdown/. Accessed 20 Feb 2014
10. Song W, Zhuang W (2009) Multi-service load sharing for resource management in the cellular/WLAN integrated network. IEEE Trans Wireless Comm 8(2):725–735
11. Zheng Z, Zhang R, Cai LX, Shen X (2012) RNP-SA: joint relay placement and subcarrier allocation in wireless communication networks with sustainable energy. IEEE Trans Wireless Comm 11(10):3818–3828
12. Jover RP (2013) Security attacks against the availability of LTE mobility networks: overview and research directions. In: Proceedings of WPMC'13, 2013
13. Dhillon HS, Ganti RK, Baccelli F, Andrews JG (2012) Modeling and analysis of Ktier downlink heterogeneous cellular networks. IEEE J Sel Area Comm 30(3):550–560
14. Ericsson (2013) LTE Release 12. In: Ericsson white paper. Ericsson. http://www.ericsson.com/res/docs/whitepapers/wp-let-release-12.pdf. Accessed 27 Feb 2014
15. Liang X, Li X, Barua M, Chen L, Lu R, Shen X, Luo H (2012) Enable pervasive healthcare through continuous remote health monitoring. IEEE Wireless Comm 19(6):10–18
16. Damnjanovic A, Montojo J, Wei Y, Ji T, Luo T, Vajapeyam M, Yoo T, Song O, Malladi D (2011) A survey on 3GPP heterogeneous networks. IEEE Wireless Comm 18(3):10–21
17. Khandekar A, Bhushan N, Tingfang J, Vanghi V (2010) LTE-advanced: heterogeneous networks. In: Proceedings of IEEE European wireless conference (EW'10), Lucca
18. Kim R, Kwak JS, Etemad K (2009) WiMAX femtocell: requirements, challenges, and solutions. IEEE Comm Mag 47(9):84–91
19. Andrews JG (2013) The seven ways HetNets are a paradigm shift. IEEE Comm Mag 51(3): 136–144
20. Yun JH, Shin KG (2011) Adaptive interference management of OFDMA femtocells for co-channel deployment. IEEE J Sel Area Comm 29(6):1225–1241
21. Andrews JG, Conti A, Shen X (2013) Best readings in multi-tier cellular. In: IEEE COMSOC best readings. IEEE Communications Society, New York. http://www.comsoc.org/best-readings/multi-tier-cellular. Accessed 10 Feb 2014
22. Cordeiro C, Challapali K, Birru D, Shankar S (2005) IEEE 802.22: the first worldwide wireless standard based on cognitive radios. In: Proceedings of IEEE DySPAN'05, pp 328–337

23. Kim H, Shin KG (2008) Efficient discovery of spectrum opportunities with MAC layer sensing in cognitive radio networks. IEEE Trans Mobile Comput 7(5):533–545
24. Akyildiz IF, Lee W-Y, Vuran MC, Mohanty S (2006) Next generation/dynamic spectrum access/cognitive radio wireless networks: a survey. Comput Netw Int J Comput Telecommun Netw 50(13):2127–2159
25. Digham FF, Alouini MS, Simon MK (2007) On the energy detection of unknown signals over fading channels. IEEE Trans Comm 55(1):3575–3579
26. Yücek T, Arslan H (2009) Survey of spectrum sensing algorithms for cognitive radio applications. IEEE Comm Surveys Tutorials 11(1):116–130
27. Krishna TV, Das A (2009) A survey on MAC protocols in OSA networks. Comput Netw (Elsevier) 53(9):1377–1394

Chapter 2
Cognitive Cellular Network Management

2.1 CCN Framework

The emerging HetNet is asking for new network deployment and management methods with flexibility to handle dynamic user demands and diverse radio environments in a tired RAN infrastructure. In this Brief, we present a new framework in the study of HetNet, namely as *cognitive cellular network management* (CCN), by applying cognitive radio techniques. As shown in Fig. 2.1, CCN are built upon four principles from bottom to top, which are spectrum awareness, effective coordination, bottleneck mitigation and integrated cellular access, respectively.

- *Spectrum Awareness* is defined as the capability of HetNet to sense and track spectrum utilization on individual cell sites and utilize the knowledge of available radio resources to fuel cellular transmissions. Since cellular communications are resource-oriented, the activities in HetNet should be aimed to improve spectrum utilization or make an easier way of improvement.
- *Effective Coordination* means that data transmissions and signaling update in HetNet should capture real time service requirements and characteristics of variations in channel/link/topology. In addition, following the cost-effective principle, network nodes should take actions to meet the designed performance requirements given available spectrum and network resources.
- *Bottleneck Mitigation* is the principle that identifies, locates and mitigates the bottlenecks in resource supply, network infrastructure and management procedures, which prohibit the performance of HetNet from further improving.
- *Integrated Cellular Access* treats HetNet problems comprehensively by identifying the corresponding roles of users, access nodes and networks in the game of service demand and resource supply. Marginal gain achieved in single technical enhancement in an access problem should be testified in a whole solution with the analysis of the paid cost.

The presented principles in CCN study are not stand-alone. They are closely interacting with each other in the research efforts to improve HetNet performance.

Y. Liu and X. Shen, *Cognitive Resource Management for Heterogeneous Cellular Networks*, SpringerBriefs in Electrical and Computer Engineering, DOI 10.1007/978-3-319-06284-6__2, © The Author(s) 2014

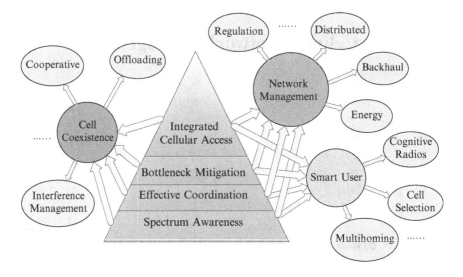

Fig. 2.1 The study framework of cognitive cellular network management (CCN)

As the foundation, spectrum awareness helps to obtain a clear picture of the resources HetNet can use, which are measured in various dimensions, e.g., time, frequency, location, and even codes. The other principles are applied on the first one, but also reward it with improved spectrum utilization. For example, effective coordination can encourage more meaningful signaling information to occupy the control bandwidth which in turn directs the network nodes to collaborate more efficiently to improve utilization of data channels. As shown in Fig. 2.1, the principles work together to addressing the challenging issues in HetNet including cell coexistence, network management and smart user. In individual research issues, different principles may take different weights according to the objective in a particular problem. In Sect. 2.2, we will present some typical applications using CCN applications and address the challenging issues in each case.

2.2 Applications and Challenges

2.2.1 Femtocell Deployment

Femtocells are a type of small cells deployed by end users to enhance the indoor cellular signal penetration in urban area with the portal to the Internet leased from third-party ISPs. Femtocells can provide the indoor public or private access to cellular users. The coexistence of femtocells with macrocells and other in-band and out-of-band small cells usually requires a spectrum access strategy because of the "near-far" problem, in which the edge macrocell users would suffer from heavy interference from the neighboring femtocells working in the same frequency band. Using spectrum awareness principle, this problem can be formulated as a prioritized spectrum access problem. Specifically, since femtocells are user-deployed, users in femtocells should yield to the priority of macrocell users in the frequency band. Analog to the similar prioritized access architecture in cognitive radio networks [1] as shown in Fig. 1.4, femtocells can coexist with macrocells under a predefined spectrum sharing method which specifies the visibility of nodes intra- and inter-user groups, priority in spectrum access and conflict resolution. Usually, the spectrum sharing methods can be categorized as overlay, underlay or interweave to agree with the requirements in different deployment scenarios. In overlay mode, for instance, macrocells actively participate in the spectrum sharing and release some bandwidth in exchange for relay assistance from femtocells to assist the transmission to edge macrocell users [2]. While in interweave mode, channel sensing and the coordination between MBS and FBS on the scheduling of resources are the primary solutions to deal with the coexistence issues since in such mode overlapping of transmissions are not allowed in the same resource blocks.

Moreover, in the tiered network infrastructure, the nodes have diverse capabilities in transmissions. The coordination between end users and cells or inter-cells greatly affects the network performance due to the mismatch of the operations or inappropriate transmission settings, which would generate severe interference. When the coordination has constraints in the network topology and limited bandwidth for the control panel, the case becomes worse. For example, the coordination between the femtocell and the macrocell is limited because the femtocell BS is indirectly connected to the cellular core network through a local Internet cable, which prohibits the operators to perform integrated network operations. Distributed decision making has proved to be a promising solution in the cognitive radio study [3]. Based upon partial and/or delayed network information, e.g., channel gains, the decision process can be modeled as a partially observable Markov decision process (POMDP) problem [4], which captures the characteristics in the interworking between the femtocells and the parent macrocell. To achieve efficient spectrum sharing among a large number of distributed users with deviated local utility functions, game theoretical approach has also been introduced into the discussions for resource management of heterogeneous cells [5].

2.2.2 Resource Management in HetNet

In HetNet, each single user senses the circumstances, e.g., channel conditions and contention level, and makes the best transmission strategy for his own utility. The egocentricity of individual transmission decisions may impair the whole network performance when effective coordination mechanisms are missing. Overall, the resource management in HetNet can be formulated as a network utility maximization problem. Specifically, under a transmission strategy, denoted by $\mathbf{a}$, which specifies the operation parameters of each node, e.g., cell selection, transmit power, etc., the objective is to maximize the aggregated utility functions of all links in the network, i.e., $\max_{\mathbf{a}} \sum_{i \in C} \sum_{j \in C_i} U_{\mathbf{a}_j}$ where C is the set of cells including all macrocells and small cells in the network, and C_i represents the set of active wireless access connections in cell i. Given the other nodes' transmissions, $\mathbf{a}_{-j}$, each node selects its transmission strategy, $\mathbf{a}_j$, to best respond to $\mathbf{a}_{-j}$, i.e., $U_{\mathbf{a}_j, \, \mathbf{a}_{-j}} \geq U_{\mathbf{a}'_j, \, \mathbf{a}_{-j}}$, $\forall \mathbf{a}_j$, $\mathbf{a}'_j \in \mathbf{a}$, $\mathbf{a}'_j \neq \mathbf{a}_j$. Furthermore, one candidate transmission strategy should not violate the network coexistence rules Γ, which determines the maximum allowable interference in the links, i.e., $\mathbf{I}_{\mathbf{a}} \leq \mathbf{I}_{\Gamma}$. The operators manipulate the decision making of individuals from the network aspect, such as load balance, interference management, and security. Candidate approaches include introducing incentive schemes [5], defining new utility functions for players [6, 7], etc.

In cellular networks, users are usually scheduled for data transmission in the time, frequency, code and space domains by a central controller. In HetNet, the centralized approach may not be available or would be costly from both computational and communication aspects. In cognitive radio networks, however, the available spectrum resources have been finely identified at different locations and times using spectrum sensing techniques [8]. The transmission pairs select the spectrum access opportunities which can satisfy the required transmission qualities, e.g., length and bandwidth. And the traffic flows are routed according to the statistics of available spectrum resources at intermediate nodes [1]. Introducing adaptive resource management in HetNet can improve the resource utilization efficiency via making opportunistic transmission decisions based on the local traffic and channel conditions.

Besides the competitions in the zero-sum game for radio resources, users and small cells can also cooperate for the channel condition monitoring, handover management and relay transmission. The cooperation can benefit the participants who have limited capability to acquire the necessary network or channel conditions to make effective decisions. No matter competition or cooperation, the participating users require the knowledge of all possible moves of other players or the required coordination information in cooperative communication. In HetNet, effective coordination relies on the connections between nodes with the overhead consideration and performance tradeoff.

2.2.3 Backhaul Bottleneck Mitigation

In HetNet, as small cells become more likely to be deployed by users themselves, it is increasingly difficult for operators to perform network resource management in a real-time manner. In addition, the capacity of access links in small cells (e.g., the links between users and SBS), and backhaul links (e.g., the ones between SBS and MBS/neighboring SBS) may vary at different paces. For example, the cellular downlink throughput can achieve 100 Mbps, while the backhaul of femtocell has limited capacity provided by the Internet service providers using digital subscriber line (DSL), normally up to 10 Mbps according to the data plan by regions and price. Therefore, the smaller bandwidth of femtocell backhaul becomes network bottleneck that limits the quality of service of users in radio access links. To tackle this problem, a possible solution is to allow multi-path data transmissions through different network interfaces, e.g., using WiFi and cellular networks [9], for the throughput aggregation at the end users.

In a HetNet where the wireless backhaul is detected as the bottleneck due to constraint link capacity, opportunistic data forwarding has proved to be an efficient solution by jointly considering the forwarding capability of femtocell BSs and the traffic loads, as proposed in [1]. Specifically, the femtocell BS evaluates its forwarding capability based upon the expected relay advancement in the forwarding direction as well as the interference in the transmission channels, which determines the order of relay candidates along the forwarding path. To fight against the fading in wireless channels, the proposed forwarding scheme incorporates multiple nodes at each transmission so that the successful receiver, if there is any, can continue with the data forwarding if the nodes with higher forwarding capability fail. Such an opportunistic forwarding scheme well adapts to the dynamic channel conditions and significantly reduces the link failures and the resulting retransmissions in the backhaul. Further detailed information of the design is presented in Chap. 3.

2.3 Research Topics

In this Brief, we will focus on two research topics, routing in wireless backhaul and interference management, preliminary works using CCN principles in the discussions are presented later in Chaps. 3 and 4, respectively.

2.3.1 Wireless Backhaul Routing

Backhaul works as a bridge for both data traffic and signaling commands commuting between the in-field SBS and the central controller/scheduler, which is critical to the success of HetNet. As shown in Fig. 2.2, base stations can be connected with each other using high speed wired (e.g., SBS3 and SBS4) or wireless links (e.g., SBS1 and SBS2). In wired backhaul, the existing fiber points of presence at macrocells can be reused to serve as the aggregation points for public access small cells. Since the deployment of these fiber POPs requires dedicate radio planning as macrocells, small cells are in many cases self deployed by users, e.g., at individual houses. In addition, fiber cannot be pulled to every lamp pole cost-effectively in many markets. Therefore, it is necessary to consider the cases when small cells are wirelessly connected to the core network, which is usually referred to as *wireless backhaul* problem in HetNet.

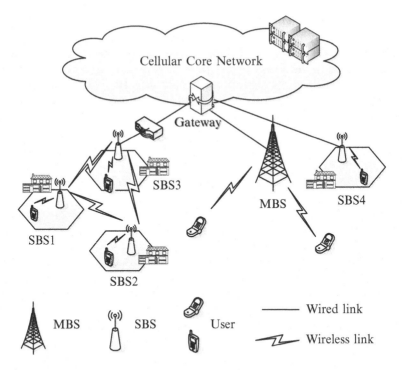

Fig. 2.2 Small cell backhaul in HetNet

Specifically, as part of the effort for a new "last 100 m", wireless backhaul is to be built out at low cost, which obtains new characteristics by cellular operators. Wireless backhaul needs to maintain both the data and control exchange over the mesh like wireless network of small cells access nodes. Existing microwave solutions using frequencies in the 6–42 GHz band cannot support discrete antennas of the kind typically required at street level [10]. These frequencies may also be running out in some cases. Therefore, the major research issue in wireless backhaul is to build the route from end SBS to the core network in a cost-effective way in terms of both spectrum and energy utilization. Given the user statistics of macrocell users in a tiered cellular access, the deployment of wireless backhaul is mainly focused on the mesh routing problem and the distributed resource management problem.

Aforementioned problem has been discussed in CRN which share the similar network model of tiered access as HetNet. The vacant UHF/VHF frequency bands for analog TV broadcasting, or "TV white spaces" (TVWS), have been proposed for wireless backhaul implementation where SBS would work as SU in CRN [11]. Routing can then be formulated as a global optimization problem with the channel-link allocation for data flows in the network [12]. Xin et al. [13] propose a layered graph to depict the topology of the SU sublayer of CRN in a snapshot and allocate multiple links over orthogonal channels to enhance the traffic throughput by establishing a near-optimal topology. Pan et al. [14] propose a joint scheduling and routing scheme according to the long term statistics of the link transmission quality for nodes. Gao et al. [15] develop a flow routing scheme which mitigates the network-wide resource for multicast sessions. These works on cognitive routing pre-determine an end-to-end relay path based on the global network information. However, the channel conditions of secondary links in wireless backhaul are highly dependent on public macrocell activities in HetNet. In addition to the limited coordination with central controller, SBS usually needs to track the channel status by periodic sensing [16] or field measurements [17]. When the channel status changes, source nodes need to re-calculate a path. Khalif et al. [18] show that the involved computation and communication overhead for re-building routing tables for all flows is nontrivial, especially when the channel status changes frequently.

Compared with centralized scheduling, distributed opportunistic routing is more suitable for the HetNet backhaul since SBS can select the next hop relay to adapt to the variations of local channel/link conditions [19, 20]. Instead of using a fixed relay path, a source node broadcasts its data to neighboring nodes, and selects a relay based on the received responses under current link conditions [19]. Liu et al. [21] propose to apply an opportunistic routing algorithm to utilize these released spectrum access opportunities in CRN where the forwarding decision is made under the locally identified spectrum opportunities. So far, most opportunistic routing protocols have been studied in a single channel scenario. In a multi-channel system, the channel selection and relay link negotiation may introduce extra delay, which degrades the performance of the network. How to extend opportunistic routing in a multi-channel HetNet is still an open research issue.

It is also recognized that with available localization services, geographic routing can achieve low complexity and high scalability under dynamic link conditions in various wireless networks, such as wireless mesh networks [22], ad hoc networks [23] and vehicle communication networks [24]. With geographic routing, a node selects a relay node that is closer to the destination for achieving distance advances in each hop. Chowdhury and Felice [25] introduce geographic routing in CRN to calculate a path with the minimal latency. However, their work still focuses on building routing tables and thus is not suitable for dynamic HetNet. Considering the unique features of HetNet, it is essential to design a distributed opportunistic routing algorithm by tightly coupling with physical layer spectrum sensing and MAC layer spectrum sharing to adapt to the network dynamics in HetNet for wireless backhaul routing.

2.3.2 Interference Management

In cognitive cellular networks, small cells are employed to enhance the link quality and improve the network capacity. When small cells operate in the same frequency band as the macrocells, severe co-channel interference degrades the performance of macrocell and small cell users. As shown in Fig. 1.2, when the mobile users served by the macrocell move to the edge of cell, they may experience strong signals from the private femtocells. Similarly, the low power transmissions in small cells are also likely interfered with macrocell users. To mitigate the co-channel interference, some candidate approaches have been proposed, including:

Spectrum splitting approach refers to the resource allocation by assigning orthogonal resources, e.g., subcarriers, to the transmission pairs with strong interference. In the tiered network, the operator can split the spectrum into subbands and assign them to the neighboring small cells to reduce the interference between the neighboring cells [26]. However, such static allocation may cause waste of spectrum and lose synchronization with the varying traffic demands.

Power control approach is to adjust the transmit power of nodes in the network to secure the reception quality at the receivers. It is a good candidate to reduce the interference in the network and encourage the energy efficient transmissions. However, the central controller needs to acquire actual channel conditions and nodes' operational parameters to optimize the performance, which introduces heavy coordination cost, especially in the tiered network infrastructure [27].

Offloading approach tries to reduce the strong interference source by arbitrarily handover these users to the cells with better link qualities to mitigate their interference over the neighbors. In this approach, both link qualities and the resource allocation needs to be considered before the handover [28]. The availability of such cell is another issue when the targeted femtocell is of closed access for its private user only.

Validation of HetNet is built on the coexistence of macrocell users and small cell users in the same frequency band, which depends on intra- and inter-cell interference management. Interference management uses a coordination mechanism among access nodes in a centralized or decentralized manner, so as to mitigate mutual interference and improve the network performance. Unlike macrocells which have been well planned before MBS deployment and managed based upon decades of research and implementation experience, the emerging small cells in HetNet are under loose control of cellular operators and have various backhaul capabilities in the coordination with macrocells and the neighboring small cells. If fibre backhaul is used to aggregate MBS and SBS links, cellular operators can centrally control the transmissions with stringent QoS as they operate macrocells. However, if the backhaul is not deterministic or with limited bandwidth, such as DSL via the Internet or wireless backhaul [11], the effective coordination and interference management are still open.

In prioritized spectrum access network, e.g., CRN, proactive and passive interference management mechanisms have been studied. In a single channel case, power control is used to maximize SUs' overall performance subject to an interference constraint at the PU side [29,30]. While in a multi-channel case, SU first choose the operating channels and then perform power control algorithms in individual channels [31,32]. Konrad et al. in [33] demonstrate that channel characteristics exhibit time-varying effects in a long time period, which is caused by the change of physical channel conditions, e.g., a light-of-sight (LOS) path between transceivers may exist for some time and disappear when the path is blocked temporarily. To make accurate estimation, the physical channel conditions should remain stable for a sufficient time to provide enough channel samples. To avoid excessive interference caused by channel uncertainty, power control methods treat the channel gain fluctuations with a stochastic model or in the worst-case approach. Zheng et al. in [34] consider the uncertain component in the channel as Gaussian noise and convert interference outage probability into a generalized Marcum's Q function [35]. In a similar way, Dall'Anese et al. in [36] approximate PUs' aggregate interference power (AIP) levels and SUs' signal to noise and interference ratios (SINRs) as log-normal distributed random variables, and then solve the resulting problem via sequential geometric programming. Gong et al. in [37] propose a robust power control method by taking the worst case calculation of the channel estimation error.

Chandrasekhar and Andrews in [38] show that the near-far problem cannot be mitigated with power control alone. Higher-layer interference management is needed, e.g. time division and spectrum splitting for mutually interfering cells, and aggressive handover from macrocell to public access small cells. Spectrum-aware MAC can facilitate small cells to find the channels with fewer active macrocell users nearby so that they can transmit at higher power for better link quality while maintaining tolerable interference on macrocell transmissions. CRN promote the MAC design by coupling the physical layer with cognitive hardware support [39], e.g., spectrum access opportunity is detected by physical layer radio frequency (RF) unit with the sensing scheduling at MAC layer, which differs from classic MAC protocols. As spectrum sensing is the key enabling functions in CRN MAC, most of the previous work mainly focus on optimal spectrum sensing policies [16, 40] or cooperative spectrum sensing among multiple SUs [41]. Kim and Shin in [16] present a sensing-period adaptation mechanism and an optimal channel sequencing algorithm. They also propose a channel usage pattern estimation technique, which can be used for efficient MAC layer scheduling. Recently, researchers pay more attention on QoS provisioning in the cognitive MAC design for real time multimedia applications. In [42], voice capacity, the maximum number of voice connections that can be supported with QoS guarantee, is analytically derived in a voice only CRN, assuming there is only one available spectrum band shared by both PUs and SUs. Kushwaha et al. in [43] distributed multimedia content over multiple unused spectrum bands based on digital fountain codes.

2.4 Summary

In this chapter, we have presented a new framework of studying HetNet, namely as CCN. Typical applications and challenges have been addressed. The research effort of wireless backhaul deployment and interference management in HetNet have been given a brief survey along with the comparative study on potential cognitive radio support. Background knowledge and literature survey on two research topics discussed in this Brief have also been presented.

References

1. Liu Y, Cai LX, Shen X (2012) Spectrum-aware opportunistic routing in multi-hop cognitive radio networks. IEEE J Select Areas Commun 30(10):1958–1967
2. Goldsmith A, Jafar SA, Maric I, Srinivasa S (2009) Breaking spectrum gridlock with cognitive radios: an information theoretic perspective. Proc IEEE 97(5):894–914
3. Saker L, Elayoubi SE, Combes R, Chahed T (2012) Optimal control of wake up mechanisms of femtocells in heterogeneous networks. IEEE J Select Areas Commun 30(3):664–672
4. Zhao Q, Tong L, Swami A, Chen Y (2007) Decentralized cognitive mac for opportunistic spectrum access in ad hoc networks: a POMDP framework. IEEE J Select Areas Commun 25(3):589–600
5. Taranto RD, Popovski P, Simeone O, Yomo H (2010) Efficient spectrum leasing via randomized silencing of secondary users. IEEE Trans Wireless Commun 9(12):3739–3749
6. Zhang J, Zhang Q (2009) Stackelberg game for utility-based cooperative cognitive radio networks. In: Proceedings of ACM MobiHoc'09
7. Zhang N, Cheng N, Lu N, Zhou H, Mark JW, Shen X (2014) Risk-aware cooperative spectrum access for multi-channel cognitive radio networks. IEEE J Select Areas Commun 32(3): 516–527
8. Yücek T, Arslan H (2009) Survey of spectrum sensing algorithms for cognitive radio applications. IEEE Commun Surv Tutor 11(1):116–130
9. Song W, Zhuang W (2009) Multi-service load sharing for resource management in the cellular/WLAN integrated network. IEEE Trans Wireless Commun 8(2):725–735
10. Patrick D (2012) Small cell backhaul: what, why and how? In: Tellabs white paper. Heavy Reading. Available: http://www.tellabs.com/resources/papers/tlab_smallcellbackhaul_wh.pdf. Accessed 27 Feb 2014
11. Small Cell Forum (2014) Doc 049.03.01 Backhaul technologies for small cells. In: Release Three: Urban Foundations. Small Cell Forum. Available: http://www.scf.io/en/documents/049_Backhaul_technologies_for_small_cells.php. Accessed 27 Feb 2014
12. Cesana M, Cuomo F, Ekici E (2011) Routing in cognitive radio networks: challenges and solutions. Ad Hoc Netw (Elsevier) 9(3):228–248
13. Xin C, Xie B, Shen C-C (2005) A novel layered graph model for topology formation and routing in dynamic spectrum access networks. In: Proceedings of IEEE DySPAN'05
14. Pan M, Zhang C, Li P, Fang Y (2011) Joint routing and link scheduling for cognitive radio networks under uncertain spectrum supply. In: Proceedings of IEEE INFOCOM'11, April 2011
15. Gao C, Shi Y, Hou YT, Sherali HD, Zhou H (2011) Multicast communications in multi-hop cognitive radio networks. IEEE J Select Areas Commun 29(4):784–793
16. Kim H, Shin KG (2008) Efficient discovery of spectrum opportunities with MAC layer sensing in cognitive radio networks. IEEE Trans Mobile Comput 7(5):533–545
17. Wellens M, Riihijarvi J, Mahonen P (2010) Evaluation of adaptive MAC-layer sensing in realistic spectrum occupancy scenearios. In: Proceedings of IEEE DySPAN'10
18. Khalif H, Malouch N, Fdida S (2009) Multihop cognitive radio networks: to route or not to route. IEEE Netw Mag 23(4):20–25
19. Zeng K, Yang Z, Lou W (2009) Location-aided opportunistic forwarding in multirate and multihop wireless networks. IEEE Trans Veh Technol 58(6):3032–3040
20. Biswas S, Morris R (2005). ExOR: opportunistic multi-hop routing for wireless networks. ACMSIGCOMM Comput Commun Rev 34(1):133–144
21. Liu Y, Cai LX, Shen X (2011) Joint channel selection and opportunistic rorwarding in multi-hop cognitive radio networks. In: Proceedings of IEEE Globecom'11
22. Karp B, Kung HT (2000) GPSR: greedy perimeter stateless routing for wireless networks. In: Proceedings of ACM Mobicom'00
23. Abdrabou A, Zhuang W (2006) A position-based QoS routing scheme for UWB Ad Hoc networks. IEEE J Select Areas Commun 24(4):850–856

24. Abdrabou A, Zhuang W (2009) Statistical QoS routing for IEEE 802.11 multihop Ad Hoc networks. IEEE Trans Wireless Commun 8(3):1542–1552
25. Chowdhury KR, Felice MD (2009) SEARCH: a routing protocol for mobile cognitive radio ad-hoc networks. Comput Commun (Elsevier) 32(18):1983–1997
26. Awad M, Mahinthan V, Mehrjoo M, Shen X, Mark JW (2010) A dual decomposition-based resource allocation for OFDMA networks with imperfect CSI. IEEE Trans Veh Technol 59(5):2394–2403
27. Almotairi KH, Shen X (2012) Distributed power control over multiple channels for ad hoc wireless networks. Wireless Commun Mobile Comput (Wiley). doi:10.1002/wcm.2266
28. Niu Z, Wu Y, Gong J, Yang Z (2010) Cell zooming for cost-efficient green cellular networks. IEEE Commun Mag 48(11):74–79
29. Gong S, Wang P, Niyato D (2010) Optimal power control in interference-limited cognitive radio networks. In: Proceedings of IEEE International Conference on Communication Systems (ICCS'10), pp 82–86, Nov 2010
30. Parsaeefard S, Sharafat A (2012) Robust worst-case interference control in underlay cognitive radio networks. IEEE Trans Veh Technol 61(10):3731–3745
31. Gong S, Chen X, Huang J, Wang P (2012) On-demand spectrum sharing by flexible time-slotted cognitive radio networks. In: Proceedings of IEEE Globecom'12
32. Chen X, Huang J (2012) Distributed spectrum access with spatial reuse. IEEE J Select Areas Commun 31(3):593–603
33. Konrad A, Zhao BY, Joseph AD, Ludwig R (2003) A Markov-based channel model algorithm for wireless networks. Wireless Netw 9(5):189–199
34. Zheng G, Ma S, Wong K-K, Ng T-S (2010) Robust beamforming in cognitive radio. IEEE Trans Wireless Commun 9(2):570–576
35. Proakis J (2000) Digital communications. McGraw-Hill, New York
36. Dall'Anese E, Kim S-J, Giannakis G, Pupolin S (2011) Power control for cognitive radio networks under channel uncertainty. IEEE Trans Wireless Commun 10(10):3541–3551
37. Gong S, Wang P, Liu Y, Zhuang W (2013) Robust power control with distribution uncertainty in cognitive radio networks. IEEE J Select Areas Commun 31(11):2397–2408
38. Chandrasekhar V, Andrews JG (2009) Uplink capacity and interference avoidance for two-tier femtocell networks. IEEE Trans Wireless Commun 8(7):3498–3509
39. Cormio C, Chowdhury KR (2009) A survey on MAC protocols for cognitive radio networks. Ad Hoc Netw 7(7):1315–1329
40. Jia J, Zhang Q, Shen X (2008) HC-MAC: a hardware-constrained cognitive MAC for efficient spectrum management. IEEE J Select Areas Commun 26(1):106–117
41. Alshamrani A, Shen X, Xie L (2009) A cooperative MAC with efficient spectrum sensing algorithm for distributed opportunistic spectrum networks. J Commun 4(10):728–740
42. Wang P, Niyato D, Jiang H (2010) Voice service capacity analysis for cognitive radio networks. IEEE Trans Veh Technol 59(4):779–1790
43. Kushwaha H, Xing Y, Chandramouli R, Heffes H (2008) Reliable multimedia transmission over cognitive radio networks using fountain codes. Proc IEEE 96(1):155–165

Chapter 3
Spectrum Aware Opportunistic Routing for Wireless Backhaul

Wireless backhaul extends small cell coverage in the "last 100 m" without wired infrastructure support, which promotes more flexible cellular penetration at low cost but obtains new characteristics that were not previously required by cellular operators [1]. Since backhaul aggregates (distributes) both data and control messages from (to) on-site cells to (from) cellular core network, wireless broadband access endorsed by adequate bandwidth is a key in the implementation of wireless backhaul as needed. However, current in-band or out-of-band solutions can not fully meet such a requirement lack of available licensed frequency bands [2, 3]. In fact, the allocated frequency bands are unevenly deployed with varying utilizations from 15 % to 85 % in time and space domains [4]. Even in crowded deployment cases, e.g., in urban area, there are still many spectrum access opportunities for unlicensed transmissions in time/frequency/space domains. A dynamic spectrum allocation mechanism which allows wireless backhaul to intelligently utilize these temporarily free channels would greatly spur the development of small cells and solve spectrum shortage in cellular networks. Small cell forum (SCF) in their latest release for best practices of HetNet has proposed to reuse TVWS for wireless backhaul in urban area where SBS would work as SUs in the vacant TV channels[3].

Utilizing spectrum access opportunities from licensed frequency bands, e.g., TVWS or cellular channels without macrocell users, SBSs in wireless backhaul form the secondary subnetwork in a CRN while PUs are the licensed users in the corresponding frequency bands. In this chapter, SBS and SU are used exchangeably hereafter. The SBS-gateway pairs in wireless backhaul may be out of the direct transmission range of each other, the multi-hop transmissions in backhaul should support opportunistic transmissions of SUs.

In this chapter, following the principle of spectrum awareness, an opportunistic cognitive routing (OCR) protocol is presented where SUs, i.e., SBSs in wireless backhaul, transmit the packets in the locally identified spectrum access opportunities. To adapt to the channel dynamics, SBS opportunistically selects the relay to or from the gateway from multiple neighboring SBSs according to the distance gain and the channel usage statistics. The main contributions of this work are fourfold: (1) In OCR, forwarding links in the route are selected based on the locally identified spectrum access opportunities for effective coordination. Specifically, the intermediate SU independently selects the next hop relay based on the local channel usage statistics so that the relay can quickly adapt to the link variations; (2) the multi-user diversity is exploited in the relay process by allowing the sender to coordinate with multiple neighboring SUs and to select the best relay node with the highest forwarding gain; (3) a novel routing metric is used to capture the unique properties of secondary transmissions, referred to as cognitive transport throughput (CTT). A heuristic algorithm is applied that achieves superior performance with reduced computation complexity based on CTT; and (4) evaluated in a multi-hop wireless backhaul simulator, simulation results show that OCR adapts well to the dynamic channel/link environment in a secondary backhaul.

3.1 System Model

We consider a wireless backhaul consisting of multiple SBSs forming a secondary network in CRN where multiple SUs (i.e., SBSs) and PUs (i.e., licensed users) share a set of orthogonal channels, $C = \{c_1, c_2, \ldots, c_m\}$. SUs can exchange messages over a common control channel (CCC).[1] Each SU is equipped with two radios: one half-duplex cognitive radio that can switch among C for data transmissions and the other half-duplex normal radio in CCC for signaling exchange.

When a source SU communicates with the gateway to the core network, which resides out of its transmission range, multi-hop relaying is required. As shown in Fig. 3.1, at each hop, the sender first senses for a spectrum access opportunity and selects a relay node in the detected idle channel.[2] We model the occupation time of PUs in each data channel as an independent and identically distributed alternating

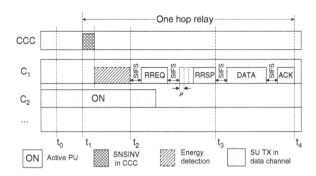

Fig. 3.1 The opportunistic cognitive routing timeline

<hr />

[1]The CCC can be implemented by bidding on a narrow spectrum band [5] or accessing the temporarily spare spectrum bands in a predefined frequency hopping sequence [6].

[2]In some extreme case when geographic routing fails to reach the destination, we can apply the right-hand rule for route recovery as proposed in GPSR [7].

ON (PU is active) and OFF (PU is inactive) process. SUs track the channel usage pattern, i.e., ON or OFF, and obtain the channel usage statistics through periodic sensing operations. Generally, the statistics of channel usage time change slowly. The parameter estimation is beyond the scope of this work and the details can be found in [8, 9]. With GPS or other available localization services, SUs can acquire their own location information, and the source nodes have the corresponding destinations' location information, e.g., an edge router or a gateway in the network. A summary of main notations used in this chapter is given in Table 3.1 for easy reference.

3.2 Spectrum Aware Opportunistic Routing

3.2.1 Protocol Overview

As shown in Fig. 3.1, the per hop relay in OCR includes three steps, i.e., channel sensing, relay selection, and data transmission.

In the channel sensing step, the sender searches for a temporarily unoccupied channel in collaboration with its neighbors using energy detection technique. Before sensing the data channel, the sender broadcasts a short message, i.e., *sensing invitation* (SNSINV), in the CCC to inform neighboring nodes of the selected data channel, and the location information of the sender and the destination. The transmission of SNSINV message in the CCC follows the CSMA/CA mechanism as specified in IEEE 802.11 MAC. Upon receiving the SNSINV, the neighboring SUs set the selected data channel as non-accessible so that no SU will transmit in the selected data channel during the sensing period of the sender. In this way, the co-channel interference from concurrent secondary transmissions can be mitigated. Using the location information in SNSINV, the neighboring SUs evaluate whether they are eligible relay candidates, e.g., whether a relay node is closer to the destination than the sender and thus can provide a relay distance gain. Eligible relay candidates will collaborate with the sender in channel sensing and relay selection. Other SUs cannot transmit in the selected data channel during the reserved time period specified in SNSINV. When the channel is sensed idle, i.e., no PU activity is detected, the sender will initiate a handshake with relay candidates in the relay selection step. Otherwise, the sender selects another channel and repeats the channel sensing process. Figure 3.2 shows a snapshot of the sender S's neighborhood. At the end of the sensing step, S detects Channel 1 is busy (occupied by a nearby PU) and selects Channel 2 to continue.

In the relay selection step, the sender selects the next hop relay from the relay candidate SUs. Specifically, when the channel is sensed idle, the sender first broadcasts a routing request (RREQ) message to the relay candidates. Eligible candidates reply routing response (RRSP) messages in a sequence specified by the sender. A relay candidate is assigned a higher priority to transmit RRSP after

Table 3.1 Summary of notations in Chap. 3

Symbol	Definition
$C = \{c_j\}$	Channel set, $j = \{1, 2, \ldots, m\}$
$\mathbf{N}_S$	The SU set of the sender S's neighbors
$\mathbf{R}_D$	The set of relay candidates for the destination D
$\mathbf{R}_D^{c_j}$	The set of relay candidates for the destination D in the channel c_j
$A_D(S, R)$	Relay advancement of the link SR for D
$(c^*, \mathbf{R}_D^{c^*})$	The transmission channel and the ordered relay set selected by MAXCTT
$CTT(c_j, \mathbf{R}_D^{c_j})$	The cognitive transport throughput (CTT)
$d(S, D)$	Euclidian distance between S and D
$E[T_{ON}^{c_j}](E[T_{OFF}^{c_j}])$	Mean duration of a busy (idle) c_j
$\mathscr{F}_{OFF}^{c_j}(t)$	CDF of the OFF duration of c_j
$I_R^{c_j} (\overline{I_R^{c_j}})$	SU R detects c_j to be idle (busy)
T_{detc}	Per channel energy detection delay
T_{DTX}	Per hop data packet transmission delay
T_{init}	Sensing initialization delay
T_{relay}	Per hop transmission delay in OCR
$T_{RREQ}(T_{RRSP})$	RREQ (RRSP) message transmission delay
T_{RS}	Per hop relay selection delay
T_{SNS}	Per hop sensing delay
T_{switch}	Transceiver switching time
t_0	The latest channel status observation time
$P_i^{c_j}$	The probability that R_i is selected as the relay in c_j
$P_{OFF,R}^{c_j}(t_0, t_1)$	The probability that c_j is idle at t_1, $t_1 > t_0$
$P_R^{c_j}(t_1, t_2)$	The probability that c_j is idle during $[t_1, t_2]$ at R
$P_{relay,R_i}^{c_j}$	The probability that the relay via R_i succeeds in c_j
$P_{RSfail}^{c_j}$	The probability that relay selection fails in c_j
V_{R_i}	The priority of R_i in the relay selection
$X_{R_t R_r}^{c_j} = 1(0)$	SU R_t and SU R_r are (not) affected by the same PU in c_j
ρ_{c_j}	The chance for an idle state in c_j
μ	Backoff mini-slot
γ	The maximum channel propagation delay

a shorter backoff window if it has a larger link throughput [10], a greater relay distance advancement [7], or a higher link reliability [11]. For example, S selects the neighbor r_2 as the relay who is closer to the destination than the other candidate, r_3, as shown in Fig. 3.2. A candidate SU keeps listening to the data channel until it overhears an RRSP or it transmits an RRSP when its backoff timer reaches zero. The sender selects the first replying relay candidate as the next hop relay. If the sender receives no RRSP message, which implies no relay candidate is available in the selected channel, it will repeat the channel sensing and relay selection steps. After a successful RREQ-RRSP handshake, the sender transmits data to the selected relay node in the data transmission step.

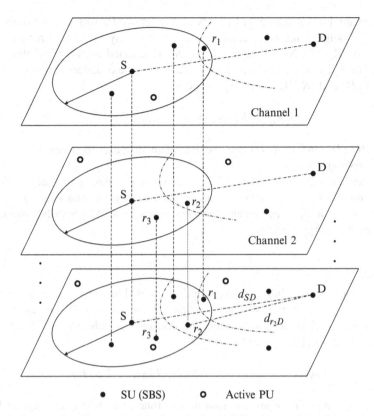

SU (SBS)   o   Active PU

Fig. 3.2 An example of location-based relay selection in OCR with two channels

3.2.2 Routing Protocol Analysis

We study the impacts of PUs' activities on the performance of the proposed OCR protocol. In CRN, when PUs appear in a channel, an SU needs to stop its current transmission, update its record of the channel status, and reselect a data channel. Thus, PUs' appearance will result in a larger transmission delay, and involve extra overhead for channel sensing and relay selection. To evaluate the impacts of PUs' activities on the protocol performance, we first introduce the main performance metrics, namely, relay distance advancement and per hop transmission delay. Based on the introduced metrics, we then analyze the success probability in each step, i.e., channel sensing, relay selection, and data transmission.

3.2.2.1 Performance Metrics

We first introduce the relay distance advancement and the per hop delay for performance evaluation. The relay advancement is measured by the geographic distance gain. For a sender S in CRN, $\mathbf{N}_S$ is the set of SUs within its transmission range. The neighboring relay candidate set for the relay to the destination D is denoted by $\mathbf{R}_D \subseteq \mathbf{N}_S$. If an SU $R \in \mathbf{N}_S$ is selected as the relay, the relay advancement $A_D(S, R)$ in terms of the difference in the distances between the SU pairs, (S, D) and (R, D) can be expressed by

$$A_D(S, R) = d(S, D) - d(R, D), \tag{3.1}$$

where $d(S, D)$ and $d(R, D)$ are the Euclidian distances between (S, D) and (R, D), respectively.

The per hop transmission delay T_{relay} is comprised of three parts: sensing delay (T_{SNS}), relay selection delay (T_{RS}), and packet transmission delay (T_{DTX}).

The sensing delay T_{SNS} includes the transmission time of an SNSINV message, T_{init}, and the energy detection time, T_{detc},

$$T_{SNS} = T_{init} + T_{detc}. \tag{3.2}$$

Based on the relay capability, candidate SUs are sorted in a given prioritized order. In the relay selection, the ith relay candidate R_i sends an RRSP message only when the first $i - 1$ higher-priority candidates are not available. Therefore, the relay selection delay $T_{RS}(i)$ is given by

$$T_{RS}(i) = T_{RREQ} + (i - 1)\mu + T_{RRSP} + 2\,SIFS, \tag{3.3}$$

where T_{RREQ} and T_{RRSP} are the transmission time of an RREQ message and an RRSP message, respectively, and μ is the duration of one mini-slot in the backoff period. According to [12], the length of a mini-slot can be calculated as $\mu = 2 \cdot \gamma + t_{switch}$, where γ is the maximum channel-propagation delay within the transmission range, and t_{switch} is the time duration that the radio switches between the receiving mode and the transmitting mode.

Once R_i is selected, the packet transmission delay T_{DTX} is

$$T_{DTX} = T_{DATA} + T_{ACK} + 2\,SIFS, \tag{3.4}$$

which includes the packet transmission delay (T_{DATA}) and the ACK transmission time (T_{ACK}).

The transmission delay $T_{relay}(R_i)$ via the relay at R_i is the delay sum

$$T_{relay}(R_i) = T_{SNS} + T_{RS}(i) + T_{DTX}. \tag{3.5}$$

3.2.2.2 Channel Sensing

Denote $I_R^{c_j}$ ($\overline{I_R^{c_j}}$) as the event that c_j is sensed to be idle (busy) by an SU R in the channel c_j. A channel is determined to be idle given that it is sensed idle at the starting time of t_1 and remains idle until sensing completes at t_2, as shown in Fig. 3.1. According to the renewal theory, the channel status can be estimated by the distribution of the channel state duration and the sensing history [13]. Specifically, given the channel status (idle or busy) observed at an earlier time, e.g., t_0, we have $P_{OFF,R}^{c_j}(t_0, t_1)$, the probability that c_j is idle (OFF) at t_1, $t_1 > t_0$. Assume ON and OFF durations follow exponential distributions with mean $1/E[T_{ON}^{c_j}]$ and $1/E[T_{OFF}^{c_j}]$.[3]

$$P_{OFF,R}^{c_j}(t_0, t_1)$$

$$= \begin{cases} \rho_{c_j} + (1 - \rho_{c_j})e^{-\Delta_{c_j}(t_1 - t_0)}, & \text{if } c_j \text{ is OFF at } t_0, \\ \rho_{c_j} - \rho_{c_j} e^{-\Delta_{c_j}(t_1 - t_0)}, & \text{if } c_j \text{ is ON at } t_0, \end{cases}$$

$$\text{where } \begin{cases} \rho_{c_j} = \dfrac{E[T_{OFF}^{c_j}]}{E[T_{ON}^{c_j}] + E[T_{OFF}^{c_j}]}, \\ \Delta_{c_j} = \dfrac{1}{E[T_{ON}^{c_j}]} + \dfrac{1}{E[T_{OFF}^{c_j}]}. \end{cases} \quad (3.6)$$

Note that ρ_{c_j} indicates the chance for an idle state in c_j.

We then calculate the likelihood of the channel staying idle during the sensing period. According to the renewal theory, the residual time of a state in an alternating process truncated since the time origin can be expressed by the equilibrium distribution of the state duration [13]. Thus, the probability that the channel at R stays in the idle state during the sensing period $[t_1, t_2]$ can be calculated as

$$P_R^{c_j}(t_1, t_2) = \int_{t_2 - t_1}^{\infty} \frac{\mathscr{F}_{OFF}^{c_j}(u)}{E[T_{OFF}^{c_j}]} du, \quad (3.7)$$

where $\dfrac{\mathscr{F}_{OFF}^{c_j}(t)}{E[T_{OFF}^{c_j}]}$ is the probability density function (PDF) of the residual time of an idle channel since the time origin when it is observed as idle. $\mathscr{F}_{OFF}^{c_j}(t)$ is the cumulative distribution function (CDF) of the duration of the OFF state in c_j with mean $E[T_{OFF}^{c_j}]$, i.e., $\mathscr{F}_{OFF}^{c_j}(t) = \int_0^t f_{T_{OFF}^k}(x)dx$. Then, the probability that R detects a spectrum access opportunity in c_j is given by

$$Pr\{I_R^{c_j}\} = P_{OFF,R}^{c_j}(t_0, t_1) \cdot P_R^{c_j}(t_1, t_2). \quad (3.8)$$

---

[3]which are commonly used in other works [8,9].

For the OCR protocol, $Pr\{I_S^{c_j}\}$ denotes the probability of sensing success when the sender S detects c_j as an idle channel. Once the sender finds an idle channel, it will move to the relay selection step. Otherwise, the sender will switch to another channel and initiate the channel sensing process.

3.2.2.3 Relay Selection

After detecting an idle channel, the sender needs to select a relay for data forwarding. In OCR, the prioritized RRSP transmission enables the relay candidate of the highest relay priority to notify the sender its availability for data forwarding. However, active PUs may interrupt the handshaking process and cause the failures in the relay selection when an SU candidate cannot reply due to the detection of active PUs. Such case is very rare, and it happens only when a nearby PU turns on during the selection period. Since the relay selection is very short in time, usually less than 1 ms, we mainly consider the case when a candidate SU detects the selected channel which is occupied by an active PU in the sensing. In this case, the candidate will not respond to the RREQ. If no relay candidate responds to the RREQ message at the moment, the relay selection fails. Therefore, we have

$$P_{RSfail}^{c_j} = Pr\{I_S^{c_j}\} \cdot Pr\left\{ \bigcap_{R_i \in \mathbf{R}_D^{c_j}} \overline{I_{R_i}^{c_j}} \middle| I_S^{c_j} \right\}, \tag{3.9}$$

where $Pr\{I_S^{c_j}\}$ indicates the probability that the sender initiates the relay selection when it detects an idle channel as defined in Eq. (3.8). In c_j, one feasible relay selection $\mathbf{R}_D^{c_j} = \{R_1, R_2, \ldots, R_n\}$ contains a set of SUs in $\mathbf{R}_D$ with the size of $n = |\mathbf{R}_D^{c_j}|$. Denote V_{R_i} as the priority of R_i in the RRSP transmission. $\mathbf{R}_D^{c_j}$ is sorted in the descending order of V_{R_i}, i.e., $V_{R_1} > V_{R_2} > \ldots > V_{R_n}$. The event that no relay candidate replies in the relay selection step, is equivalent to the event that all SUs in $\mathbf{R}_D^{c_j}$ sense the channel busy in the previous sensing with the probability $Pr\left\{ \bigcap_{R_i \in \mathbf{R}_D^{c_j}} \overline{I_{R_i}^{c_j}} \middle| I_S^{c_j} \right\}$.

In the CRN, we assume that an SU is affected by at most one active PU in one frequency band. Such assumption holds in the frequency bands such as the downstream bands in cellular networks where the adjacent cells/sectors are usually assigned with different working frequencies to avoid the co-channel interference [14]. Thus, the channel usage pattern is mainly determined by the PU activity at the spot of the individual SU. Let $X_{R_t R_r}^{c_j} = 1$ if a pair of SUs, R_t and R_r, are affected by the same PU in c_j, and $X_{R_t R_r}^{c_j} = 0$ otherwise. $X_{R_t R_r}^{c_j}$ can be acquired and maintained by the periodic exchange of the channel status in the SU's neighborhood. A cognitive transmission is successful only if both ends of the link are not influenced by active PUs. For example, if the channel utilities of c_j at R_t and R_r are $\rho_{R_t}^{c_j}$ and $\rho_{R_r}^{c_j}$, respectively, the link quality of the link l_{tr} can be expressed by $P_{l_{tr}}^{c_j} = \rho_{R_t}^{c_j} \cdot \rho_{R_r}^{c_j \, (1-X_{R_t R_r}^{c_j})}$. Therefore, $Pr\left\{ \bigcap_{R_i \in \mathbf{R}_D^{c_j}} \overline{I_{R_i}^{c_j}} \middle| I_S^{c_j} \right\}$ in Eq. (3.9) is given by

$$Pr\left\{ \bigcap_{R_i \in \mathbf{R}_D^{c_j}} \overline{I_{R_i}^{c_j}} \middle| I_S^{c_j} \right\}$$

$$= Pr\left\{ \overline{I_{R_1}^{c_j}} \middle| I_S^{c_j} \right\} \cdot \prod_{i=2}^{n} Pr\left\{ \overline{I_{R_i}^{c_j}} \middle| \left\{ \bigcap_{k=1}^{i-1} \overline{I_{R_k}^{c_j}} \right\} \cap I_S^{c_j} \right\}$$

$$= (1 - X_{SR_1}^{c_j}) Pr\left\{ \overline{I_{R_1}^{c_j}} \right\}$$

$$\cdot \prod_{i=2}^{n} \left[(1 - X_{SR_i}^{c_j}) Pr\left\{ \overline{I_{R_i}^{c_j}} \right\}^{\prod_{k=1}^{i-1}(1 - X_{R_k R_i}^{c_j})} \right]. \tag{3.10}$$

Suppose that the ith relay candidate R_i in the selected relay selection order $\mathbf{R}_D^{c_j}$ is available, R_i will be selected as the next hop relay with the probability $P_i^{c_j}$, given that previous $i - 1$ candidates are not available,

$$P_i^{c_j} = \begin{cases} Pr\{I_S^{c_j}\} \cdot Pr\{I_{R_1}^{c_j} | I_S^{c_j}\}, & \text{for } i = 1, \\[4mm] Pr\{I_S^{c_j}\} \cdot Pr\left\{ \left\{ \bigcap_{k=1}^{i-1} \overline{I_{R_k}^{c_j}} \right\} \cap \{I_{R_i}^{c_j}\} \middle| I_S^{c_j} \right\}, \\[4mm] & \text{for } 2 \le i \le n, \end{cases} \tag{3.11}$$

where $Pr\left\{ \left\{ \bigcap_{k=1}^{i-1} \overline{I_{R_k}^{c_j}} \right\} \cap \{I_{R_i}^{c_j}\} \middle| I_S^{c_j} \right\}$ can be expressed as

$$Pr\left\{ \left\{ \bigcap_{k=1}^{i-1} \overline{I_{R_k}^{c_j}} \right\} \cap \{I_{R_i}^{c_j}\} \middle| I_S^{c_j} \right\}$$

$$= Pr\left\{ \overline{I_{R_1}^{c_j}} \middle| I_S^{c_j} \right\} \cdot \prod_{u=2}^{i-1} Pr\left\{ \overline{I_{R_u}^{c_j}} \middle| \left\{ \bigcap_{r=1}^{u-1} \overline{I_{R_k}^{c_j}} \right\} \cap I_S^{c_j} \right\}$$

$$\cdot Pr\left\{ I_{R_i}^{c_j} \middle| \left\{ \bigcap_{k=1}^{i-1} \overline{I_{R_k}^{c_j}} \right\} \cap I_S^{c_j} \right\}$$

$$= (1 - X_{SR_1}^{c_j}) Pr\left\{ \overline{I_{R_1}^{c_j}} \right\}$$

$$\cdot \prod_{u=2}^{i-1} \left[(1 - X_{SR_i}^{c_j}) Pr\left\{ \overline{I_{R_u}^{c_j}} \right\}^{\prod_{r=1}^{u-1}(1 - X_{R_r R_u}^{c_j})} \right]$$

$$\cdot \left[\prod_{k=1}^{i-1} (1 - X_{R_k R_i}^{c_j}) \right] Pr\{I_{R_i}^{c_j}\}^{(1 - X_{SR_i}^{c_j})}. \tag{3.12}$$

3.2.2.4 Data Transmission

Once R_i is selected, the data transmission in the link l_{SR_i} succeeds when no active PU appears during the transmission period $[t_3, t_4]$ in c_j. Thus, the successful relay probability at current hop via R_i can be expressed by

$$
\begin{aligned}
P_{relay,R_i}^{c_j} &= P_i^{c_j} \cdot P_{l_{SR_i}}^{c_j}(t_3, t_4) \\
&= P_i^{c_j} \cdot P_S^{c_j}(t_3, t_4) \cdot P_{R_i}^{c_j}(t_3, t_4)^{(1-X_{SR_i}^{c_j})}.
\end{aligned} \tag{3.13}
$$

3.3 Joint Channel and Relay Selection

We then jointly consider the selection of the sensing channel and relay node to improve the performance of the proposed OCR. As many factors, including channel usage statistics, the relay distance advances, and transmission priority of relay candidates, may affect the relay performance, we introduce a new metric to capture these factors and apply it in a heuristic algorithm to select the best relay in one data channel at a reduced computation complexity.

3.3.1 Novel Routing Metric

We design a new metric, the CTT, $CTT(c_j, \mathbf{R}_D^{c_j})$, to characterize the one hop relay performance of OCR in the selected channel c_j with the selected relay candidate set $\mathbf{R}_D^{c_j}$, in unit of bit·meter/second.

$$
\begin{aligned}
CTT(c_j, \mathbf{R}_D^{c_j}) &= E\Big[L \cdot \frac{A_D^{c_j}}{T_{relay}^{c_j}}\Big] \\
&= \sum_{R_i \in \mathbf{R}_D^{c_j}} P_{relay,R_i}^{c_j} \frac{L \cdot A_D(S, R_i)}{T_{relay}(R_i)}
\end{aligned} \tag{3.14}
$$

The physical meaning of the CTT defined in Eq. (3.14) is the expected bit advancement per second for one hop relay of a packet with the payload L in the channel c_j. To improve the OCR performance, we should maximize the one hop relay performance along the path as one hop performance improvement contributes to the end-to-end performance. In addition, as the multi-user diversity is implicitly incorporated in the relay selection process, we can also achieve a high multi-user diversity gain by maximizing CTT. From Eq. (3.14), we can jointly decide channel c_j and the corresponding relay selection order $\mathbf{R}_D^{c_j}$ to maximize CTT.

3.3.2 Heuristic Algorithm

To obtain c^* and $\mathbf{R}_D^{c^*}$ for the largest CTT, we can exhaustively search for all possible combinations of the sensing channel and the subset of the relay candidate set. Given m channels and up to n relay candidates, an exhaustive search needs to find the locally optimal one in each channel by comparing the value of CTT under all possible permutations of the set of relay candidates. Since the CTT value is sensitive to the set size as well as the permutation, given that k candidate nodes are incorporated in the relay selection, $1 \leq k \leq n$, there are $P(n,k)$ types of opportunistic forwarding patterns. Therefore, over m channels, the exhaustive search should take $m \cdot \sum_{k=1}^{n} P(n,k)$ times of the CTT calculation to return the global optimum. If n goes to infinity, we can get $\lim_{n \to \infty} m \cdot \sum_{k=1}^{n} P(n,k) = \lim_{n \to \infty} m \cdot \sum_{k=0}^{n} \frac{n!}{(n-k)!} = \lim_{n \to \infty} m \cdot n! \cdot \left[\sum_{k=0}^{n} \frac{1}{k!} - 1 \right]$. Thus the exhaustive search running time is $O(m \cdot n! \cdot e)$, where e is the base for natural logarithms. We can see that once n becomes very large, the exhaustive search becomes infeasible in real implementations.

To reduce the complexity, we propose an efficient heuristic algorithm to reduce the searching space yet achieve similar performance of the optimal solution. The performance comparison will be given in the following section.

Given independent channel usage statistics in different channels, we can decompose the optimization problem into two phases. First, we compare all possible relay selection orders in each channel and find the optimal one which maximizes the CTT. Then, we choose the relay selection order with the largest CTT value over all channels and select the corresponding channel as the sensing channel. Since the number of channels is usually limited, it is more important to reduce the searching complexity for the best relay selection order in a single channel.

To find the optimal relay selection order, the sender should decide both the number of the relay candidates and the relay priority of each candidate. According to Eq. (3.14), a neighboring SU, R_i, is an eligible relay candidate if it contributes to a positive relay distance advancement, $A_D(S, R_i)$. One feasible relay selection order $\mathbf{R}_D^{c_j}$ in c_j is an ordered subset of $\mathbf{R}_D$ in the descending order of relay priority V_{R_i}. A larger size of $\mathbf{R}_D^{c_j}$ include more relay candidates and achieves a higher diversity gain, which improves the per hop throughput at the cost of increased searching complexity.

To reduce the searching space and improve the algorithm efficiency, we have the following lemma.

Lemma 3.1. *Given a feasible relay selection set $\mathbf{R}_D^{c_j}$, $\exists R_{i_1}, R_{i_2} \in \mathbf{R}_D^{c_j}$, if $V_{R_{i_1}} > V_{R_{i_2}}$, $X_{R_{i_1} R_{i_2}}^{c_j} = 1$, then $CTT(c_j, \mathbf{R}_D^{c_j} \setminus \{R_{i_2}\}) \geq CTT(c_j, \mathbf{R}_D^{c_j})$.*

Proof. Suppose $\mathbf{R}_D^{c_j} = \{R_1, \ldots, R_{i_1}, \ldots, R_{i_2}, \ldots\}$. According to Eq. (3.11), if $V_{R_{i_1}} > V_{R_{i_2}}$, $X_{R_{i_1} R_{i_2}}^{c_j} = 1$ and $X_{R_{i_1} R_{i_2}}^{c_j} = 1$, $P_{i_2}^{c_j} = 0$. Thus, $P_{relay, R_{i_2}}^{c_j} = 0$. From Eq. (3.14),

$$CTT(c_j, \mathbf{R}_D^{c_j}) = \sum_{r=1}^{i_2-1} P_{relay, R_r}^{c_j} \frac{L \cdot A_D(S, R_r)}{T_{relay}(R_r)}$$

$$+ \sum_{r=i_2+1}^{|\mathbf{R}_D^{c_j}|} P_{relay, R_r}^{c_j} \frac{L \cdot A_D(S, R_r)}{T_{relay}(R_r)}$$

$$\leq \sum_{r=1}^{i_2-1} P_{relay, R_r}^{c_j} \frac{L \cdot A_D(S, R_r)}{T_{relay}(R_r)}$$

$$+ \sum_{r=i_2+1}^{|\mathbf{R}_D^{c_j}|} P_{relay, R_r}^{c_j} \frac{L \cdot A_D(S, R_r)}{T_{relay}(R_r) - \mu}$$

$$= CTT(c_j, \mathbf{R}_D^{c_j} \setminus \{R_{i_2}\}),$$

which shows that the CTT performance does not drop when R_{i_2} is deleted from $\mathbf{R}_D^{c_j}$.

Lemma 3.1 indicates that we can reduce the size of the relay selection by excluding the relay candidates that are affected by the same PU. The reduced set of relay candidates will not degrade CTT. Specifically, for a given set of relay candidates, the sender groups the SUs that are affected by the same PU, selects the SU with the highest relay priority, and deletes other SUs in a group from the set.

We observe the following property which can be used to further reduce the searching space.

Property 3.1 (Tail Truncation Rule). Given a feasible relay selection $\mathbf{R}_D^{c_j}$, $\exists R_i \in \mathbf{R}_D^{c_j}$, $X_{SR_i}^{c_j} = 1$, then $CTT(c_j, \mathbf{R}_D^{c_j}) = CTT(c_j, \mathbf{R}_D^{c_j} \setminus \{R_k | R_k \in \mathbf{R}_D^{c_j}, V_{R_k} < V_{R_i}\})$.

Proof. If S and R_i are affected by the same PU, $Pr\{\overline{I_{R_i}^{c_j}} | I_S^{c_j}\} = 0$. According to Eq. (3.11), $P_k^{c_j} = 0, \forall R_k \in \mathbf{R}_D^{c_j}, V_{R_k} < V_{R_i}$. Thus,

$$CTT(c_j, \mathbf{R}_D^{c_j}) = \sum_{r=1}^{i} P_{relay, R_r}^{c_j} \frac{L \cdot A_D(S, R_r)}{T_{relay}(R_r)}$$

$$+ \sum_{r=i+1}^{|\mathbf{R}_D^{c_j}|} 0 \cdot \frac{L \cdot A_D(S, R_r)}{T_{relay}(R_r)}$$

$$= CTT(c_j, \mathbf{R}_D^{c_j} \setminus \{R_k | R_k \in \mathbf{R}_D^{c_j},$$

$$V_{R_k} < V_{R_i}\}),$$

which shows that the CTT performance does not change when the relay candidates are removed from $\mathbf{R}_D^{c_j}$ with lower priority than R_i.

Property 3.1 indicates that the size of the relay candidate set can be further reduced by deleting SUs whose relay priorities are lower than the SU that is affected by the same PU as the sender. In other words, we can reduce the searching set without degrading the performance of the current flow while the deleted candidates can also participate in other transmissions, which further improve the network performance.

As discussed above, the relay priority plays a critical role in relay selection. It is well known that in geographic routing, users closest to the destination is the best next hop relay as it provides the greatest distance gain. It is also proved that the geographic routing approaches the shortest path routing with the distance advance metric [15]. Therefore, we also apply the distance advance and verify its efficiency in the proposed OCR.

Thus, the CTT metric can be approximated as

$$CTT(c_j, \mathbf{R}_D^{c_j}) \simeq \frac{L}{T_{relay}} \cdot \sum_{i=1}^{|\mathbf{R}_D^{c_j}|} P_{relay,R_i} A_D(S, R_i)$$

$$= \frac{L}{T_{relay}} \cdot E[A_D^{c_j}], \tag{3.15}$$

where $E[A_D^{c_j}]$ is the estimated relay advancement in c_j, and T_{relay} is the estimated one hop transmission delay in Eq. (3.5). To maximize the CTT in each channel, we need to find an optimal relay selection to maximize $E[A_D^{c_j}]$. When opportunistic routing over independent links uses $E[A_D^{c_j}]$ as a routing metric, [15] has proved that the optimal relay priority should be set according to the distance of the relay candidate to the destination. In addition, the maximum $E[A_D^{c_j}]$ increases with the number of relay candidates. Therefore, we can assign the relay priority in the descending order of $A_D(S, R)$.

We then propose a heuristic algorithm, MAXCTT, as shown in Algorithm 1. The inputs are the channel set C, the set of relay candidates $\mathbf{R}_D$, and the maximum number of relay candidates in relay selection r_{max}. MAXCTT selects the SUs from $\mathbf{R}_D$ to form the relay selection order $\mathbf{R}_D^{c_j}$ and calculates the achieved CTT_{c_j} in each c_j. By comparing CTT_{c_j} over the channels, MAXCTT returns the channel c^* that has CTT_{max} and the corresponding relay selection order $\mathbf{R}_D^{c^*}$ as the algorithm output.

Specifically, an eligible relay candidate set $\mathbf{R}_E$ is first formed by excluding the SUs affected by the same PU in c_j according to Lemma 3.1, which is a subset of $\mathbf{R}_D$ (line 4–line 6). A recursive searching [10] is then applied to obtain $\mathbf{R}_D^{c_j}$ (line 8–line 14). At the beginning of the searching step, $\mathbf{R}_D^{c_j}$ contains no SU. Each time, $\mathbf{R}_D^{c_j}$ includes one more relay candidate out of the remaining SUs in $\mathbf{R}_E$ which provides the best CTT improvement. The selected relay candidates are sorted in the descending order of $A_D(S, R)$ in $\mathbf{R}_D^{c_j}$. The formed $\mathbf{R}_D^{c_j}$ contains all relay candidates

---

Algorithm 1: The MAXCTT algorithm

Input: the channel set C, the relay candidate set $\mathbf{R}_D$, r_{max}
Output: the selected channel c^*, the relay selection order $\mathbf{R}_D^{c^*}$

1: $c^* \leftarrow 0$; $\mathbf{R}_D^{c^*} \leftarrow \emptyset$; $CTT_{max} \leftarrow 0$;
2: **for** each c_j **do**
3:     $\mathbf{N} \leftarrow \mathbf{R}_D$; $\mathbf{R}_E \leftarrow \emptyset$; $\mathbf{R}_D^{c_j} \leftarrow \emptyset$; $R_p \leftarrow \emptyset$; $CTT_{c_j} \leftarrow 0$;
4:     **while** $(\mathbf{N} \neq \emptyset)$ **do**
5:         $\mathbf{R}_E \leftarrow$ insert an SU $R_i \in \mathbf{N}$ that has max $A_D(S, R_i)$;
            Remove $R_j \in \mathbf{N}$ with $X_{R_i R_j} = 1$ from $\mathbf{N}$;
6:     **end while**
7:     **while** $(\mathbf{R}_E \neq \emptyset$ && $|\mathbf{R}_D^{c_j}| < r_{max}$ && $X_{SR_p} \neq 1)$ **do**
8:         **for** each SU $R_i \in \mathbf{R}_E$ **do**
9:             $\mathbf{R}_T \leftarrow \mathbf{R}_D^{c_j} + R_i$; Sort $\mathbf{R}_T$ in the descending order of $A_D(S, R)$;
                Get CTT on $\mathbf{R}_T$ according to Eq. (3.14);
10:            **if** $(CTT > CTT_{c_j})$ **then**
11:                $CTT_{c_j} \leftarrow CTT$; $R_p \leftarrow R_i$;
12:            **end if**
13:        **end for**
14:        $\mathbf{R}_D^{c_j} \leftarrow$ insert R_p in the descending order of $A_D(S, R)$; $\mathbf{R}_E \leftarrow \mathbf{R}_E - R_p$;
15:    **end while**
16:    **if** $(CTT_{c_j} > CTT_{max})$ **then**
17:        $c^* \leftarrow c_j$; $\mathbf{R}_D^{c^*} \leftarrow \mathbf{R}_D^{c_j}$; $CTT_{max} \leftarrow CTT_{c_j}$;
18:    **end if**
19: **end for**
20: **return** $(c^*, \mathbf{R}_D^{c^*})$;

---

from $\mathbf{R}_E$, and it satisfies the requirements of r_{max} and Property 3.1 (line 8). The search ends when (1) all relay candidates are included, or (2) the set size reaches the upper boundary, i.e, r_{max}, or (3) the relay selection needs to be truncated according to Property 3.1. The recursive searching obtains the optimal $\mathbf{R}_D^{c_j}$ in c_j when the size of the selection order is at most 2, and it achieves almost the same performance as the optimal solution when the final order contains more than two candidates according to Lemma 5.1 in [10]. Suppose that the largest size of $\mathbf{R}_E$ over the channels is n, at most $m \cdot \sum_{k=1}^{n} k$ times of the CTT calculations are required to find CTT_{max}. Thus, the time complexity of MAXCTT is $O(m \cdot n^2)$, which is much lower than exhaustive search.

3.4   Simulation Results

In this section, we evaluate the performance of the OCR protocol by simulation under different network settings, e.g., channel conditions, number of SUs, and traffic loads, using an event-driven simulator coded in C/C++ [16, 17]. The network parameter settings are shown in Table 3.2 if no other specification is made in the individual study.

Table 3.2 Simulation parameters in performance evaluation of Chap. 3

Number of channels	6
$\{\rho_{c_1}, \rho_{c_2}, \rho_{c_3}, \rho_{c_4}, \rho_{c_5}, \rho_{c_6}\}$	$\{0.3, 0.3, 0.5, 0.5, 0.7, 0.7\}$
Number of PUs per channel	11
PU coverage	250 m
$E[T_{OFF}]$	[100 ms, 600 ms]
Number of SUs	[100, 200]
SU transmission range	120 m
Source-destination distance	700 m
SU CCC rate	512 kbps
SU data channel rate	2 Mbps
CBR delay threshold	2 s
Mini-slot time, μ	4 μs
Per channel sensing time	5 ms
Channel switching time	80 μs
PHY header	192 μs
r_{max}	2

3.4.1 Simulation Settings

The PU activity in each channel is modeled as an exponential ON–OFF process with parameters $1/E[T_{ON}]$ and $1/E[T_{OFF}]$, and the idle rate $\rho = E[T_{OFF}]/(E[T_{ON}] + E[T_{OFF}])$ is selected accordingly. The channel status is updated by periodic sensing and on-demand sensing before data transmissions. We set up a CRN with multiple PUs and a wireless backhaul consisting of SBSs as SUs, all of which are randomly distributed in an $800 \times 800 \text{ m}^2$ area. We set a pair of SUs as the source SBS and the backhaul gateway with a distance of 700 m, and a constant bit rate (CBR) flow is associated with the SU pair with packet size 512 bytes and flow rate of 10 packets per second (pps). The unit disc model is applied for the data transmission. The channel switch time is 80 μs [18], the minimum sensing duration with energy detection is 5 ms, and a mini-slot is 4 μs [12]. We evaluate the performance of the proposed OCR protocol in terms of the end-to-end delay, the packet delivery ratio (PDR) and the hop count, i.e., the total number of transmission hops between the source and destination SUs. We run each experiment for 40 s and repeat it 500 times to calculate the average value.

We then compare the performance of the OCR protocol with that of SEARCH [19], based on different metrics for the channel and relay selection, which are listed as follows.

1. SEARCH: SEARCH [19] is a representative geographic routing protocol in tiered network infrastructure. It sets up a route with the minimal latency before data transmissions. If an active PU is detected which blocks the route, SEARCH pauses the transmissions and recalculates the route. We modify SEARCH by updating route periodically to adapt to the dynamic changing spectrum access opportunities along the route.

2. OCR (CTT): For OCR (CTT), the channel and the relay candidate set are jointly selected by using the proposed CTT metric and heuristic algorithm proposed in Sect. 3.3.
3. OCR (OPT): For OCR (OPT), the channel and the relay candidate set are determined by exhaustively searching for the biggest CTT over all possible channel-relay sets.
4. GOR: For geographic opportunistic routing (GOR) algorithm, the SU first selects the channel with the greatest success probability of packet transmissions; if the channel is sensed idle, the SU then select a relay SU over the channel. The relay selection order is based on the location information and the relay capability of SUs [20].
5. GR: For geographic routing (GR), an SU first selects the channel for sensing as in GOR. If the selected channel is sensed idle, the SU then selects the SU closest to the destination as the next hop relay.

3.4.2 PU Activities

We first evaluate the performance of OCR under different PU activity patterns. The average PU OFF duration $E[T_{OFF}]$ varies from 100 ms (high channel dynamics) to 600 ms (low channel dynamics). The PDR performance of OCR and SEARCH are compared under different traffic loads in Fig. 3.3a. A smaller $E[T_{OFF}]$, e.g., 100 ms, indicates the available time window is shorter and thus SUs' transmissions are more likely to be interrupted by PUs. We can see a marked PDR improvement under dynamic channel conditions for the per hop relay schemes, e.g., OCR (CTT), compared with SEARCH which is based on the global route establishment. In OCR (CTT), SUs are allowed to locally search and exploit spare spectrum and select the available links to data forwarding. Thus, OCR (CTT) can adapts well in the dynamic data channels. On the contrary, SEARCH uses a pre-determined routing table. Once an active PU is detected along the relay path, intermediate SUs should defer the packet relay until they update their routing tables according to the current channel availabilities in CRN. Since more SUs are involved in the route establishment, the handshakes between SUs in the network to establish the relay path introduce a large overhead and results in a longer delay.

Figures 3.3b and 3.4 compare the end-to-end delay performance. All routing protocols achieve a better delay performance when the idle channel state becomes longer, e.g., from 100 ms to 600 ms, as more packets can be transmitted during the idle state. When the channel state change frequently, SEARCH needs to update routing tables accordingly which involves a long delay for route recovery. Our proposed OCR protocols are opportunistic routing algorithms that quickly adapt to the dynamic channel environment and achieve better delay performance compared with SEARCH. OCR (CTT) also outperforms GR and GOR since the latter two protocols perform the channel and relay selection separately while OCR (CTT) jointly consider the channel selection and relay selection.

Fig. 3.3 Performance comparison between OCR and SEARCH under different channel conditions (Number of SUs: 200, flow rate: 10/40 pps) (**a**) Packet delivery ratio (PDR) (**b**) End-to-end delay

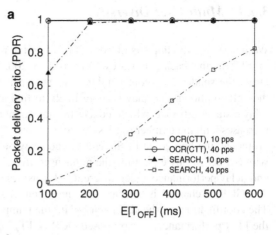

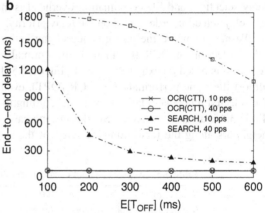

Fig. 3.4 Performance comparison under different traffic loads and PU activities (Number of SUs: 200, flow rate: 10 pps)

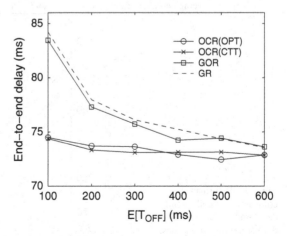

3.4.3 Multi-User Diversity

We investigate the impacts of node density on the relay performance. The number of SUs in the backhaul varies from 100 to 200. When the number of SUs is large, the sender has more neighbors as shown in Table 3.3. With more SUs in the neighborhood, the relay is more likely to find a feasible relay link with better relay distance advance, which reduces the hop count number. The relay performance increases with the number of SUs due to the larger diversity gain. As a result, for all protocols, the hop count of the end-to-end relay decreases and the PDR increases with SU density by exploiting the multi-user diversity in the backhaul. The end-to-end delay performance under different SU densities is compared in Fig. 3.5. For GR and GOR, a channel is selected first, and then SUs coordinates to serve as relay. The coordination overheads increase with the number of SUs, which also degrades the PDR performance. The proposed OCR (CTT) jointly considers the channel and relay selection, and SU coordination overhead is minimized as sender determines the relay selection order based on the relay priority.

We next compare the performance of the heuristic algorithm for the channel-relay selection in OCR (CTT) with the optimal one in OCR (OPT) where the selection is based on exhaustive search. Figure 3.5 shows that OCR (CTT) achieves almost the same performance as OCR (OPT), even when the returned number of the selected relay candidates is over 2, according to the value of r_{max} in Table 3.2. Table 3.3 indicates that as the SU density increases in the network, the number of neighbors along the forwarding direction of the sender will increase accordingly.

Table 3.3 Average neighbor density under different SU densities

Number of SUs	100	120	140
Average number of neighbors	7.0686	8.4823	9.8960
Number of SUs	160	180	200
Average number of neighbors	11.3097	12.7235	14.1372

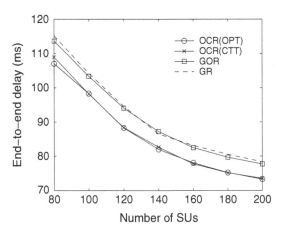

Fig. 3.5 Performance comparison of end-to-end delay under different SU densities (flow rate: 10 pps, $E[T_{OFF}] = 200$ ms)

For example, given 160 SUs over 6 channels, the average number of neighbors of an SU is around 11. OCR (OPT) takes over 6.5×10^8 times of the CTT calculation to find the globally optimal solution which is infeasible for real time implementation. In the simulated scenario, although at most four neighbors are under independent PU coverage which significantly reduces the searching space, OCR (OPT) still takes 384 runs while OCR (CTT) only needs 60 runs, which achieves the marked reduction at the computational expense.

3.4.4 Effectiveness of Routing Metric

We further compare the performance of OCR (CTT) with that of GR and GOR to evaluate the effectiveness of routing metrics used in the channel and relay selection. We first compare the performance under different traffic loads. We change the traffic load by varying the flow rate from 10 pps (light load) to 70 pps (heavy load). As shown in Fig. 3.6a, b, when the traffic load increases, the PDR and delay

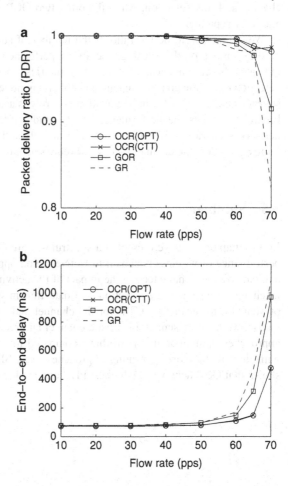

Fig. 3.6 Performance comparison under different traffic loads (Number of SUs: 200, $E[T_{OFF}] = 200$ ms) (**a**) Packet delivery ratio (PDR) (**b**) End-to-end delay

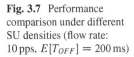

Fig. 3.7 Performance comparison under different SU densities (flow rate: 10 pps, $E[T_{OFF}] = 200$ ms)

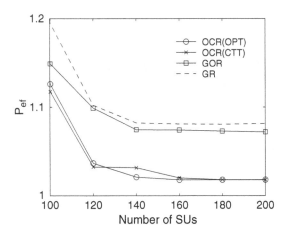

performance degrade. However, the decreasing rate of OCR (CTT) is much lower than that of GR and GOR. This is because OCR (CTT) jointly considers the optimal channel and link selection, while the other two OCR protocols select the channel and relay separately.

We define P_{ef} to be the ratio of the number of successful relay transmissions to the number of the sensing operations performed in the data channels. P_{ef} indicates the effectiveness of the routing metrics since the transmission relies on detection of an idle channel and an available relay node. If P_{ef} approaches to 1, the selected channel for each hop relay almost surely is available for data transmission. Figure 3.7 shows the performance of P_{ef} under different node densities. In all network scenarios, OCR (CTT) outperforms GR and GOR, because CTT metric jointly considers the channel access and relay selection.

3.5 Summary

In this chapter, we have presented a spectrum-aware OCR protocol to improve the performance of wireless backhaul in HetNet when opportunistic spectrum reuse is applied. We have characterized the impact of PU activities on the operation of OCR in channel sensing, relay selection and data transmission. Furthermore, we have presented a novel metric, CTT, for the channel and relay selection. Based on the metric, we have presented a heuristic channel and relay selection algorithm which approaches optimal solution. We have compared the performance of OCR (CTT) with that of the existing routing approaches, e.g., SEARCH, GR and GOR and shown that OCR achieves the highest PDR and the lowest delay.

References

1. Patrick D (2012) Small cell backhaul: what, why and how?. In: Tellabs white paper. Heavy Reading. Available: http://www.tellabs.com/resources/papers/tlab_smallcellbackhaul_wh.pdf. Accessed 27 Feb 2014
2. NTIA (2003) U.S. frequency allocations. Available: http://www.ntia.doc.gov/osmhome/allochrt.pdf. Accessed 20 Feb 2014
3. Small Cell Forum (2014) Doc 049.03.01 Backhaul technologies for small cells. In: Release Three: Urban Foundations. Small Cell Forum. Available: http://www.scf.io/en/documents/049_Backhaul_technologies_for_small_cells.php. Accessed 27 Feb 2014
4. FCC (2003) ET Docket No. 03-222 Notice of proposed rule making and order.
5. Pefkianakis I, Wong SHY, Lu S (2008) SAMER: spectrum aware mesh routing in cognitive radio networks. In: Proceedings of IEEE DySPAN'08
6. Timmers M, Pollin S, Dejonghe A, Van der Perre L, Catthoor F (2010) Distributed multichannel MAC protocol for multihop cognitive radio networks. IEEE Trans Veh Technol 59(1):446–459
7. Karp B, Kung HT (2000) GPSR: greedy perimeter stateless routing for wireless networks. In: Proceedings of ACM Mobicom'00
8. Wellens M, Riihijarvi J, Mahonen P (2010) Evaluation of adaptive MAC-layer sensing in realistic spectrum occupancy scenearios. In: Proceedings of IEEE DySPAN'10
9. Kim H, Shin KG (2008) Efficient discovery of spectrum opportunities with MAClayer sensing in cognitive radio networks. IEEE Trans Mobile Comput 7(5):533–545
10. Zeng K, Yang Z, Lou W (2009) Location-aided opportunistic forwarding in multirate and multihop wireless networks. IEEE Trans Veh Technol 58(6):3032–3040
11. Khalife H, Ahuja S, Malouch N, Krunz M (2008) Probabilistic path selection in opportunistic cognitive radio networks. In: Proceedings of IEEE Globecom'08, pp 4861–4865
12. Bi Y, Cai LX, Shen X, Zhao H (2010) Efficient and reliable broadcast in inter-vehicle communications networks: a cross layer approach. IEEE Trans Veh Technol 59(5):2404–2417
13. Cox DR (1967) Renewal theory. Butler and Tanner, London
14. Castro JP (2001) The UMTS network and radio access technology. Wiley, New York
15. Zeng K, Lou W, Yang J, Brown DR (2007) On geographic collaborative forwarding in wireless ad hoc and sensor networks. In: Proceedings of WASA'07, Chicago, IL, Aug 2007
16. Liu Y, Cai LX, Shen X, Mark JW (2011) Exploiting heterogeneity wireless channels for opportunistic routing in dynamic spectrum access networks. In: Proceedings of IEEE ICC'11
17. Chigan C (2012) Cognitive radio cognitive network simulator. Available: http://faculty.uml.edu/Tricia_Chigan/Research/CRCN_Simulator.htm. Accessed 27 Feb 2014
18. Sampath A, Yang L, Cao L, Zheng H, Zhao BY (2008) High throughput spectrum-aware routing for cognitive radio based ad hoc networks. In: Proceedings of CROWNCOM'08, May 2008
19. Chowdhury KR, Felice MD (2009) SEARCH: a routing protocol for mobile cognitive radio ad-hoc networks. Comput Commun (Elsevier) 32(18):1983–1997
20. Liu Y, Cai LX, Shen X (2011) Joint channel selection and opportunistic forwarding in multihop cognitive radio networks. In: Proceedings of IEEE GLOBECOM'11

Chapter 4
QoS-Aware Cognitive MAC and Interference Management for HetNet

Validation of HetNet is built on the coexistence of macrocells and small cells in the same frequency band, which depends on intra- and inter-cell interference management. Chandrasekhar and Andrews in [1] have shown that the near-far problem cannot be fully mitigated with power control alone. Higher-layer interference management is needed, e.g. time division and spectrum splitting for mutually interfering cells, and aggressive handover from macrocell to public access small cells. In addition, compared with "public users" served by macrocells, users in self-deployed small cells, e.g., private femtocells, usually have lower access priority in the available channels assigned by the network operators. To meet their respective traffic demands, small cell users can search for adequate channel bandwidth in a more opportunistic way. Users can access the free portion of different frequency bands according to the agreement of secondary access, e.g., in TVWS [2], or using unlicensed bands to transmit [3]. Based on frequency hopping, dynamic channel access can feed user-deployed small cells with the increasing data hungry cellular services and diverse deployment needs.

In this chapter, we present a distributed QoS-aware MAC scheme which facilitates users in private access small cells to find the channels with fewer active macrocell users nearby so that they can transmit at higher power for better link quality while maintaining tolerable interference on macrocell transmissions. Specifically, based on the knowledge of channel usage patterns of macrocell users, users in small cells sense a set of channels for available bandwidth for various types of traffic. The MAC function is further enhanced in QoS provisioning by applying differentiated ASPs to satisfy their QoS requirements in support of multimedia applications. An analytical model is then developed to study the MAC layer performance taking the activities of both macrocell users and small cell users into consideration. Along with the dynamic channel access, a penalty based power allocation scheme is used to manipulate the transmissions of small cell users in a cost-effective way given limited backhaul bandwidth. Extensive simulations validate the analysis and demonstrate that the presented MAC can achieve multiple levels of QoS provisioning for various types of multimedia applications in HetNet.

Y. Liu and X. Shen, *Cognitive Resource Management for Heterogeneous Cellular Networks*, SpringerBriefs in Electrical and Computer Engineering, DOI 10.1007/978-3-319-06284-6_4, © The Author(s) 2014

4.1   System Model

4.1.1   Network Model

We assume that a HetNet consists of single hop small cells where all private users can directly communicate with their SBSs in a distributed manner. The HetNet has Nc data channels and one common control channel. As a dedicated universal spectrum band may not be always available in small cells, ultra-wideband (UWB) is considered as an ideal option for unlicensed common control channel [4].[1] Exploiting the spreading technique in UWB system, senders use their receivers' spreading codes to initiate handshake transmissions over the unlicensed UWB control channel. After a successful handshake, or namely transceiver synchronization, both the sender and receiver tune to the same data channel for communications.

Each small cell is modeled as a CRN in which a number of private users opportunistically access the data channels without causing harmful interference to the users in macrocells. Private users in small cells (public users in macrocells) and SUs (PUs) are used exchangeably in this chapter. Due to the hardware constraints, SUs can only sense one channel at a time. We assume an SU can accurately determine the channel status after a basic sensing period, e.g., 1 ms. Each data channel is modeled as an alternating ON (PU is active)/ OFF (PU is inactive) model. We assume the channel parameters are slowly time varying and each SU is able to track the channel usage pattern by PUs through periodic sensing. How to estimate the variations of the channel usage and determine the parameters of ON/OFF periods is out of the scope of this Brief and more details can be found in [5].

4.1.2   Traffic Model

With the ever-increasing demand of wireless multimedia services, e.g, voice of IP (VoIP) and video conference, SUs may carry various applications, such as voice, video, and data, which have different QoS requirements in terms of throughput and delay. In this study, we consider multiple services supported in HetNet.

For voice traffic, we consider VoIP application here. In a VoIP system, the sampled voice signals in analogue are formed into constant-bit-rate (CBR) flows after the compression and encoding. For example, iLBC voice codec is a very bandwidth efficient codec with a rate of 15.2 kbps and has been used in Internet soft-phone applications, e.g., Skype. The voice traffic is modeled as a CBR traffic flow with a packet interval of 20 ms and a packet size of 38 Bytes [6]. To satisfy the QoS demand for voice traffic, according to international telecommunication union

---

[1]UWB is especially attractive as it is power limited and causes little interference to other communication networks.

(ITU) standards [7], the one-way end-to-end delay of voice traffic should be no greater than 150 ms for good voice quality and up to 400 ms for acceptable voice quality, with an echo canceler. Also, the voice packet loss rate should be no more than 1 % to maintain satisfactory voice quality.

Compared with voice traffic, realtime video traffic usually demands higher throughput and is modeled as a variable bit rate (VBR) flow with different compression ratios and various payload formats in the codec. In this study, we consider the codec of H.263 and H.264/MPEG-4 AVC, which support very efficient video compression and are applied in a broad range of video applications from low rate Internet video streaming to high definition video (HDV), such as Flash video contents as used on sites as YouTube, Google Video. For example, a H.263 video "Star Trek - First Contact" trace file has the mean rate of 256 kbps and the peak rate of 1.5 Mbps [8]. The mean frame rate and frame size are 25 frame per second and 4, 420 Bytes, respectively. The general QoS metrics of video applications include throughput, delay, jitter, and packet loss.For interactive applications, e.g., video telephony, the normal tolerable delay should be less than 100 ms and packet loss rate should be below 1 %.

We also consider the background data service which is modeled as a simple saturated data flow in the performance study of CRN. Currently, there are many data applications in the Internet services including email, web browsing, file transfer, etc. Each SU is assumed to have a data packet in the queue ready for transmission. The saturated data model is applicable for large volume bulk data transfer applications. Although data traffic is usually delay-insensitive, generally it requires no transmission error. Transmission errors can be improved by the schemes applied in the link layer and the reliable transmission control protocols in the transport layer. In this chapter, we evaluate the QoS metrics for data traffic in terms of throughput and fairness.

4.2 QoS-Aware Cognitive MAC for Small Cells

We present a cognitive MAC protocol with QoS provisioning for multimedia services in small cells of Hetnet. By exploiting diverse channel usage patterns, SUs in small cells adapt to PUs' activities in macrocells and select appropriate channels that satisfy their QoS requirements for channel sensing and data transmissions. The transmission procedure of an SU is shown in Fig. 4.1. Without loss of generality, an SU senses the first channel and starts data transmission if the channel is sensed idle for a sensing interval. To reduce possible collisions among SUs, each SU will sense the channel for an ASP, which consists of the basic sensing period that assures satisfactory sensing accuracy plus some random slots selected from a sensing window $[0, SW_i]$. If the channel is sensed busy, SU switches to the second channel. The pseudo code of the SU transmission procedure is shown in Fig. 4.1. Note that at the beginning of the channel sensing period, the sender will initiate a handshake with its receiver over the control channel for transceiver synchronization.

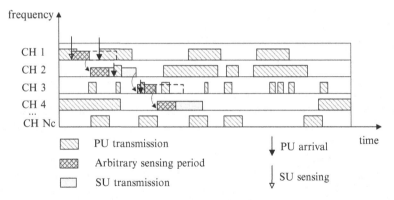

Fig. 4.1 Multi-channel transmissions in small cells

It is also possible that a PU may appear during SU's data transmission, in which case the transmission fails and the SU will switch to the next channel to retransmit the data. Therefore, the key research issue in the cognitive MAC design is how to effectively explore the spectrum opportunities and select an appropriate set of channels to assure the QoS performance of SUs.

4.2.1   Channel Sensing

Denote the transmission time of the current frame of user i as t_i, which is constant for voice traffic but may vary for video traffic. To fully utilize the spectrum opportunity, SUs calculate the probability that the current frame can be successfully transmitted over channel k, which is given by

$$P_i^k = Pr(\text{Channel k is idle, Transmission is successful})$$

$$= \frac{\beta_k}{\alpha_k + \beta_k} F_k(t_{AS_i} + t_i), \tag{4.1}$$

where $1/\alpha_k$ and $1/\beta_k$ are the mean idle and busy periods of channel k, t_{AS_i} is an ASP selected by user i, t_i is the transmission time of the current frame, and $F_k(x)$ is the cumulative distribution function $F_k(x) = \int_0^x f_k(x)dx$ that characterizes the usage pattern of channel k by the PU's activity, the derivation of which can be found in [5]. Notice that with an ASP, it is less likely that two or more SUs transmit simultaneously and collide with each other. Therefore, the transmission is successful if no PU appears during the following $ASP_i + t_i$ interval.

Based on the calculated successful transmission probability in each channel, SUs can determine the channel sensing sequence with two different policies: greedy and ascending. For the greedy policy, SUs simply sort channels in a descending order

---

Algorithm 2: SU transmission procedure in small cells

Begin:
  1: **if** (User i has a data for User j) **then**
  2:    Calculate P_s^i and decide the sensing sequence;
  3:    Select the first channel;
  4:    Sense the channel and initiate a handshake with User j over the control channel;
  5:    **if** (Channel is sensed busy during ASP_i **then**
  6:      Select the next channel;
  7:      Go to line 4;
  8:    **else**
  9:      Transmit data;
10:      **if** (Transmission is successful) **then**
11:        Go to End;
12:      **else**
13:        Go to line 6;
14:      **end if**
15:    **end if**
16: **end if**
End

---

and always use channels with the highest success probability for achieving a low delay and high throughput. However, the channel with less PUs' activity is more likely to be selected by SUs, which causes high contention level among SUs sharing the same radio resources and degrade the performance accordingly, as shown in algorithm 2. Therefore, we propose the second sensing policy that allows different SUs select various channels based on the QoS requirements of their applications. For instance, each realtime frame is associated with the maximum tolerable one hop delay τ_i. Given P_i^k, the expected transmission time over channel k is estimated as $E[T_k^i] = (t_{AS_i} + t_i)/P_i^k$. Therefore, an SU first selects a group of channels that satisfy

$$E[T_i^k] = (t_{AS_i} + t_i)/P_i^k < \tau_i, \tag{4.2}$$

and senses these channels in the ascending order of P_i^k. That is, each SU first selects a channel with the minimum P_k^i that satisfies its delay requirement. If the channel is sensed busy, the SU will switch to the next channel with second lowest P_k^i. As the channel is sorted in the ascending order of P_k^i, the expected delay of SU i, $E[T_i] \leq E[T_k^i] < \tau_i$. Therefore, the delay performance can be guaranteed with the ascending policy. Notice that although SU can estimate the channel usage pattern by PUs, it is difficult if not impossible for an SU to accurately estimate the number of SUs currently sharing the spectrum bands. For a simple yet robust MAC design, an SU can set a stringent delay bound τ_i and select a channel sets with more opportunities to incorporate the impacts of other SUs. The impacts of contention level among SUs will be analytically studied in Sect. 4.2.3, which can provide important guideline for an SU to set the parameter τ_i and select a set of channels for opportunistic transmissions.

Table 4.1 Sensing window design for multimedia services

	Strict priority	Statistical priority	No priority
Voice	[0,31]	[0,31]	[0,31]
Video	[32,63]	[0,63]	[0,31]
Data	[64,127]	[0,127]	[0,31]

4.2.2   Service Differentiation

We further enhance the QoS provisioning of the proposed cognitive MAC by introducing service differentiation in the ASPs of different traffic flows. Basically, a smaller sensing window is applied for a higher priority real time applications so that they have a higher chance to access data channels when opportunity appears, i.e., $SW_{voice} < SW_{video} < SW_{data}$. In addition, by carefully determining the sensing windows for different types of traffic, multiple levels of QoS provisioning can be achieved for multimedia applications in CR networks. As shown in Table 4.1, a statistical priority is provided by simply doubling the sensing windows for various types of traffic, while a strict priority can be achieved when non-overlapped sensing windows are used. The performance of the differentiated service provisioning using different settings will be evaluated in Sect. 4.4.

4.2.3   Performance Analysis

In this section, we develop an analytical model to study the delay performance of the proposed cognitive MAC.

An SU senses channel k and attempts to transmit if the channel is sensed idle for t_{AS_i}. In other words, an SU's sensing fails if (1) the channel has been occupied by a PU with probability

$$P_{CO}^k = Pr(\text{Channel occupied}) = \frac{\alpha_k}{\alpha_k + \beta_k}; \qquad (4.3)$$

(2) the channel is idle but a PU turns on during the sensing interval with probability

$$P_{PU}^k = Pr(\text{PU on}) = \frac{\beta_k}{\alpha_k + \beta_k} F_k(t_{AS_i}) \qquad (4.4)$$

or (3) the channel becomes busy due to any other SU's transmissions. We consider a homogeneous case that all SUs use a constant sensing window for channel access. Let the maximum sensing window $SW_i = W$. Given there are N_k SUs contending in channel k, the probability that the tagged SU wins the contention, i.e., all of the remaining SUs select a larger sensing window than the tagged SU, is given by

$$Pr(\text{SU } i \text{ wins contention}) = \sum_{j=1}^{W} \frac{1}{W} (\frac{W - j}{W})^{N_k}. \qquad (4.5)$$

As the SUs contend for channel access only when no PU activity is detected, the probability that an SU's sensing fails due to other SU's transmission is given by

$$P_{SU}^k = \frac{\beta_k}{\alpha_k + \beta_k}(1 - F_k(t_{AS_i}))(1 - \sum_{j=1}^{W}\frac{1}{W}(\frac{W-j}{W})^{N_k}). \tag{4.6}$$

As it is very complicated to track the number of SUs in each data channel due to highly dynamic spectrum access in the channels, we use the average number of SUs to estimate the contention level in each channel, $N_k = (N-1)/N_d$, where N is the total number of SUs in the system, N_d is the number of data channels selected by the SU for transmissions. Therefore, the probability that an SU succeeds in sensing and attempts a transmission over channel k is

$$P_{ss}^k = Pr(\text{sensing succeeds}) = 1 - P_{CO}^k - P_{PU}^k - P_{SU}^k. \tag{4.7}$$

An SU transmits data when its sensing succeeds, or it switches to the next channel when sensing fails. The average time an SU spends on one transmission over channel k is

$$E[T^k] = P_{ss}^k(t_{AS_i} + t_i) + (1 - P_{ss}^k)t_{AS_i}. \tag{4.8}$$

A transmission succeeds only if no PU turns on during the total sensing and transmission time of the SU,

$$P_{TS}^k = Pr(\text{No PU transmit and SU } i \text{ wins})$$

$$= \frac{\beta_k}{\alpha_k + \beta_k}\sum_{j=1}^{W}\frac{1}{W}(\frac{W-j}{W})^{N_k}F_k(t_{AS_i} + t_i). \tag{4.9}$$

Or the transmission fails due to the disruption from PU with the probability

$$P_{TF}^k = P_{ss}^k F_k(t_i). \tag{4.10}$$

Without loss of generality, an SU checks the set of selected channels in a round robin sequence, $\{CH_1, CH_2, \ldots, CH_{N_d}, CH_1, \ldots\}$, until the packet is successfully transmitted. The probability that a transmission succeeds in the rth attempts is

$$Ps(r) = P_{TS}^r \prod_{j=1}^{r-1}(1 - P_{TS}^j), \tag{4.11}$$

where P_{TS}^r corresponds to the probability of a successful transmission over the channel in the rth attempts. We obtain the average transmission delay of an SU as

$$E[T] = \sum_{r=1}^{\infty} E[T^r]P_{TS}^r \prod_{j=1}^{r-1}(1 - P_{TS}^j). \tag{4.12}$$

4.3 Power Allocation Under Violation Penalty

The QoS-aware cognitive MAC guides users in small cells to reside in the channels with less interference to macrocell transmissions. However, in each single channel, there would be many active small cell transmissions in the entire HetNet, the accumulated interference at individual macrocell users would be notable to impairs the network performance without power allocation. As shown in Fig. 4.2, Femtocell 2 is located at the edge of the macrocell. In the downlink, the leaked signal from Femtocell 2 to the nearby macrocell user, UE2, may be stronger than UE2's received signal from the macrocell base station, which is referred to as the "near-far" problem, especially when UE2 is located at the edge of the macrocell. Therefore, Femtocell 2 introduces significant interference on UE2's transmission. Many literatures on femtocells have addressed such problem [9]. However, existing solutions mainly focus on the centralized resource management, which may not be suitable for a tiered cognitive cellular network where a robust distributed approach is more desirable due to the random deployment of private access small cells, i.e., femtocells. As a case study, we present an effective power allocation scheme among femtocells with limited coordination bandwidth in backhaul.

4.3.1 Effective Control in Constrained Backhaul

In practical operations, the central controller of the macrocell can hardly fully control the affiliated femtocells because these femtocells may not follow the scheduling information but like to aggressively compete for network resources to maximize their own utility. For example, femtocell base stations can increase their

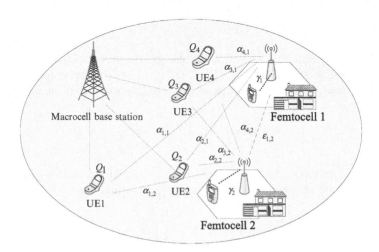

Fig. 4.2 Interference management in RAN with femtocell deployment

transmit power for achieving higher throughput while causing greater interference to the neighboring users. In HetNet, the central controller may not be able to specify the violation behaviors of individual femtocells, even when the neighboring macrocell users report the reception failure caused by such violation. In other words, the central controller can not effectively to eliminate the co-interference resulting from self-deployed femtocells. To analyze the motivation and behavior of femtocells in the violation of the centralized scheduling, we apply a game theory approach to study power management, which is widely used for resource allocation among PUs and SUs in cognitive radio networks [10]. We derive the downlink interference, and the uplink analysis can be obtained in a similar way.

Figure 4.2 shows a network snapshot in the HetNet where a group of private access femtocells with active connections coexist with macrocell users in a single downlink channel, i.e, used for transmissions from MBS (SBS) to users. In the macrocell, there are M active macrocell users in the downlink, denoted by $U = \{u_1, u_2, \ldots, u_M\}$. u_i has a threshold Q_i, which indicates the maximum tolerable interference level in the downlink. The macrocell users are scheduled to transmit in non-overlapping resource blocks so that there is no interference among macrocell users. The active femtocells in the downlink form a set $F = \{f_1, f_2, \ldots, f_N\}$ with the size of N. As indicated by the design of QoS-aware cognitive MAC in Sect. 4.2, in each channel, FBS schedules at most one single user to transmit at one time. Therefore, there is one active link in each femtocell at any time. The transmit power of the femtocell f_j is denoted by P_j. We assume that the channel gains, $\alpha_{i,j}$ of the link between u_i and f_j, $\epsilon_{j,k}$ of the link between f_j and f_k, and γ_j of the link within f_j are known to each femtocell as well as Q_i of each u_i, and the links are symmetrical.

The aggregated interference at u_i should satisfy $P_1\alpha_{i,1} + P_2\alpha_{i,2} + \cdots + P_N\alpha_{i,N} \le Q_i$, otherwise, u_i is blocked. The capacity function c_j for femtocell f_j is defined as

$$c_j = log_2(1 + \frac{P_j\gamma_j}{N_0 + \Sigma_{k \ne j} P_k\gamma_k}) \tag{4.13}$$

It is obvious that FBS prefers to using the maximal transmission power to achieve the highest link capacity. Therefore, femtocells may like to violate the power control strategy made by the central controller, which may cause co-channel interference with macrocell users. To address this issue, we apply randomized silencing policy proposed in cognitive radio networks [11]. The policy is very straightforward, i.e., if any macrocell user u_i experiences the interference greater than its limit Q_i, the central controller will randomly select one active femtocell from F and force it to turn off in current transmission period. Such silencing process continues for several rounds until no macrocell user reports the block case.[2]

---

[2]The shutdown process is valid in cellular networks where macrocell users are protected with higher priority because they have been admitted in the serving macrocell. When the self deployed femtocells register at the cellular operator, they are required to yield to the priority of macrocells if conflictions occur.

4.3.2   Game Theoretic Power Allocation

Given a power allocation strategy of femtocells, $\mathbf{P} = \{P_1, P_2, \ldots, P_N\}$, once the interference requirement is met at each macrocell user, the utility of macrocell is determined. Therefore, the power control problem in this case study can be formulated as

$$\max_{\mathbf{P}} \sum_{j \in F} E[c_j \cdot \mathbf{1}_j] \tag{4.14}$$

$$\text{s.t.}\ \ P_1 \alpha_{i,1} + P_2 \alpha_{i,2} + \cdots + P_N \alpha_{i,N} \leq Q_i, \forall i \in U$$

$$\mathbf{P} \in NE$$

where the objective of resource allocation is to find the maximum aggregated utility of femtocells, which can be denoted as $\max_{\mathbf{P}} \sum_{j \in F} E[c_j \cdot \mathbf{1}_j]$ with the function $\mathbf{1}_j = 1$ if f_j is not shut down after the silencing process, and 0 otherwise. As each femtocell intends to maximize its utility by selecting the transmission power best responding to the transmission powers of other nodes, a candidate $\mathbf{P}$ would be a power allocation of Nash equilibrium (NE).

Here, we present some preliminary results to explore the NEs for the optimal value. Using the theorems in [11], we can easily prove the orthogonal power allocation $\mathbf{P}_{OR} = \{\min_i\{\frac{Q_i}{\alpha_{i,1}}\}, \min_i\{\frac{Q_i}{\alpha_{i,2}}\}, \ldots, \min_i\{\frac{Q_i}{\alpha_{i,N}}\}\}, i \in U$ is an NE.

Moreover, we notice that a macrocell user causes significant interference when it is closer to some active femtocells than others. Therefore, the femtocell needs to avoid transmission in the resource block assigned to the nearby macrocell user if it is detected by the femtocell.[3] The femtocell can learn the allocation of resource blocks of each macrocell user by listening to the allocation message broadcasted by the macrocells at the beginning of each transmission period. In such case, we can also prove that the orthogonal power allocation $\mathbf{P}_{ORT} = \{\min_{i \in U \setminus S_1}\{\frac{Q_i}{\alpha_{i,1}}\}, \min_{i \in U \setminus S_2}\{\frac{Q_i}{\alpha_{i,2}}\}, \ldots, \min_{i \in U \setminus S_N}\{\frac{Q_i}{\alpha_{i,N}}\}\}$ is an NE where S_j is the set of macrocell users near f_j with the channel gain $\alpha_{i,j}$ greater than a predefined threshold.[4]

---

[3]It is valid in cellular networks because the users measure the signal strength from visible cells and the femtocell can detect such nearby macrocell user through this process although it may rejects the user's access request if the user is not a private member.

[4]Comparable with macrocell users, femtocells have also the predetermined QoS threshold. For example, in a femtocell f_j, it has the SINR threshold μ_j, then the selected transmission power should be greater than $\frac{\mu_j}{\gamma_j}$ for successful reception at the receiver.

4.4 Simulation Results

4.4.1 Simulation Settings

In this chapter, we evaluate the performance of the QoS-aware cognitive (QC) MAC via extensive simulations written in C. Three types of flows are simulated in a single hop small cell, i.e., voice, video, and data, as described in Sect. 4.1. The initial arrival time of each flow is uniformly distributed in [0, 5 ms]. The channels are modeled as exponential ON/OFF models with the parameters listed in Table 4.2. The capacity of each channel is 10 Mbps. The channel switch time plus the basic sensing duration is 1 ms. SUs add an arbitrary number of mini-slots after the basic sensing duration and each mini-slot is 4 μs. The arbitrary sensing window setting is tabulated in Table 4.2. The delay bound of realtime voice and video traffic is set to be 20 ms. We run each experiment for 100s and repeat them 50 times to calculate the average value.

4.4.2 Delay of Homogeneous Traffic

We first study the delay performance of the MAC in support of homogeneous traffic, i.e., voice or video flows, in Fig. 4.3. We compare the voice and video delay performance of the QC MAC with that of fractional (FRC) scheme which senses the channel in the descending order of the average channel available time. All SUs use the same sensing window [0, 31] without service differentiation. It can be seen that the average delay of voice/video traffic increases with the number of SUs. The delay of voice packets using both QC and FRC schemes are low because small voice packets are more likely to be transmitted opportunistically when PUs are inactive. For video traffic with much larger payloads, the probability of transmission failure becomes high as a PU is more likely to turn on and interfere with the SU during a longer transmission time of a video packet. When a transmission fails, an SU will switch to the next channel for sensing and retransmission, which results in a longer delay. It is also shown in Fig. 4.3 that the proposed QC MAC achieves much lower delay compared with FRC. This is because, in QC MAC, SUs always select a proper

Table 4.2 Settings of channel usage patterns in performance evaluation of Chap. 4

Channel	α	β	$\frac{\alpha}{\alpha+\beta}$	Channel	α	β	$\frac{\alpha}{\alpha+\beta}$
CH1	0.215	0.4	0.351	CH6	0.1	0.1	0.5
CH2	0.054	0.1	0.35	CH7	0.653	0.4	0.62
CH3	0.278	0.4	0.41	CH8	0.163	0.1	0.62
CH4	0.069	0.1	0.409	CH9	1.2	0.4	0.75
CH5	0.4	0.4	0.5	CH10	0.3	0.1	0.75

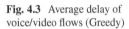

Fig. 4.3 Average delay of voice/video flows (Greedy)

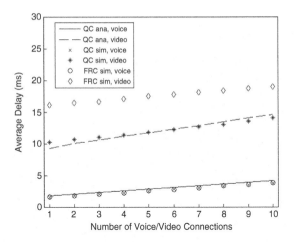

set of channels that assure high probability of successful frame transmissions, while only the average channel utilization is considered in FRC. As shown in the figure, the analytical results approximate those obtained by simulations well.

4.4.3 Delay of Heterogeneous Traffic

We then study the performance of QC MAC supporting heterogeneous traffic in the small cell with different channel sensing policies, i.e., greedy and ascending, under various traffic loads in Figs. 4.4 and 4.5. When traffic load is low, e.g., there are two video flows and one to five voice flows in the network, greedy scheme achieves better delay performance than ascending scheme. Using the greedy scheme, SUs always select the channels with the highest success probabilities so that channels with good condition, e.g., fewer PU activities, will be efficiently utilized. When more voice flows joins in the network, ascending scheme slightly outperforms greedy for video traffic. For the greedy scheme, all SUs are likely to select the best channels for their transmissions and the contention level becomes high in those channels as the number of SU increases. In the ascending scheme, different users may select various channels that satisfy their QoS requirements, and thus the contentions among SUs are distributed over multiple channels. As shown in Fig. 4.5, when there are ten video flows and up to 30 voice flows, ascending scheme achieves better delay performance for video traffic. As small voice packets are more likely to take any opportunity for transmission and less likely to be interrupted by PUs, the delay performance of voice flows are low in all cases. Overall, greedy scheme is suitable for CRN with light traffic loads, while ascending scheme is more efficient when there are multiple types of SUs with different QoS requirements. Our proposed schemes achieves a much lower delay than FRC under different traffic loads.

Fig. 4.4 Comparison
of channel sensing policies
(low traffic load)

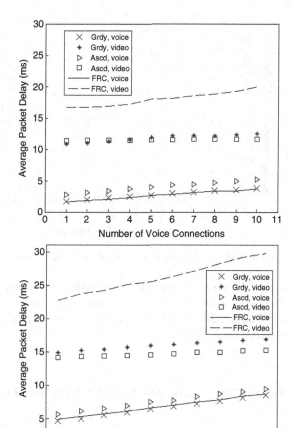

Fig. 4.5 Comparison
of channel sensing policies
(high traffic load)

4.4.4   Performance of Service Differentiation

We also investigate in the performance of the service differentiation scheme, using the sensing window setting listed in Table 4.2. We have ten voice and ten video flows in the network. To study the impacts of background data transmissions, a saturated data flow is set up in each channel. As shown in Fig. 4.6, the delay of voice traffic does not change much with the traffic loads in the network; the delay of video traffic slightly increases; while the data throughput decreases when more video SUs join in the network. By applying different sensing windows for voice, video, and data, multimedia traffic have a higher priority to be transmitted when opportunity appears. It can be seen that the voice delay is around 7 ms using strict sensing window setting, 9 ms when statistical window is applied, and 16 ms for constant window setting. Similarly, the video delay is around 27–30, 30–34, and 50–55 ms for strict, statistical, and constant sensing window settings, respectively. When a strict priority setting is applied, data packets have a lower probability to

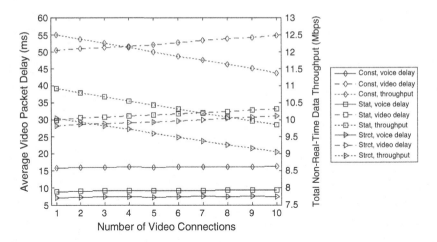

Fig. 4.6 Performance comparison with different sensing windows

access the channel, and thus multimedia applications achieves a better delay at the cost of a lower throughput of data flows. Given different types of multimedia applications, differentiated service is required to provision QoS for delay-sensitive real time applications.

4.4.5   Power Allocation Under Violation Penalty

To evaluate the performance of the presented power allocation scheme in Sect. 4.3, we simulate the network with Rayleigh fading channels where all $\alpha_{i,j}$ fade independently with average $\bar{\alpha}$, all $\epsilon_{j,k}$ fade independently with average $\bar{\epsilon}$, and all γ_j fade independently with average $\bar{\gamma}$ [11]. We set $M = 20$ and $N = 5$ in the macrocell, and $Q = 1$, $\bar{\alpha} = 1$ dB, $\bar{\epsilon} = 1$ dB, and $\bar{\gamma}$ ranges from 1 to 50 dB. In each experiment, we randomly select two femtocells with a significant neighboring macrocell user, i.e, $\bar{\alpha} \approx \bar{\gamma}$. It can be seen in Fig. 4.7 that the recognition of the significant interference sources in femtocells can improve the interference management in cellular network and achieve a higher throughput for femtocells.

4.5   Summary

In this chapter, we have presented a distributed QoS-aware MAC with service differentiation for small cells of HetNet supporting heterogeneous multimedia applications. An analytical model has been developed to study the QoS performance of the QC MAC, considering the activities of both macrocell users and small cell users. Based upon the discussion above, we have discussed the effective power

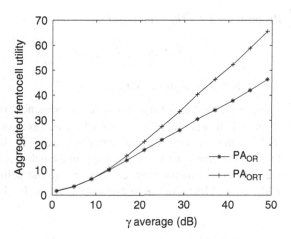

Fig. 4.7 Aggregated femtocell utility under intra-femtocell channel gains

allocation in the interference management problem. On the network bottleneck mitigation, regarding the limited bandwidth from the femtocell to the core network, we have designed the solution using game theory to analyze the nodal behavior in favor of its own utility. On the coordination mechanism design, we have developed the random shutdown scheme as the penalty for misbehavior which generates less control overhead while maintaining effective regulation effect on small cells. Simulation results have validated the analysis, and shown that the QC MAC provides satisfactory QoS performance for multimedia applications.

Appendix: Proof of Nash Equilibrium in Sect. 4.3

Proposition 4.1. $\mathbf{P}_{OR} = \{\min_i\{\frac{Q_i}{\alpha_{i,1}}\}, \min_i\{\frac{Q_i}{\alpha_{i,2}}\}, \ldots, \min_i\{\frac{Q_i}{\alpha_{i,N}}\}\}, i \in U$ is an NE under random silencing.

Proof. With orthogonal power allocation, each femtocell user generates an interference at each macro user equal to Q. According to the violation penalty, if any macro user reports the interference violation, the random silence will proceed. Therefore, from the point view of each single femtocell user, it only has the chance to transmit with the power equal to the minimum allowed power by any active macro user nearby, i.e., $\min_i\{\frac{Q_i}{\alpha_{i,j}}\}$ for femtocell user j. On the one hand, no user can increase its utility by decreasing its power. On the other hand, if any user increases its power (thus violating the Q), then it is shut down with certainty and its utility is always zero. This shows that $\mathbf{P}_{OR}$ is always a NE.

Next, if the most significant interference sources have been allocated in orthogonal resource slots, then, for each femtocell, it can increase its instant transmission utility by increasing the power in the active resource slots. However, the total resource slots have been reduced.

Proposition 4.2. *The orthogonal power allocation*

$$\mathbf{P}_{ORT} = \{ \min_{i \in U \setminus S_1} \{\frac{Q_i}{\alpha_{i,1}}\}, \min_{i \in U \setminus S_2} \{\frac{Q_i}{\alpha_{i,2}}\}, \ldots, \min_{i \in U \setminus S_N} \{\frac{Q_i}{\alpha_{i,N}}\}\}$$

is a NE under random silencing.

Proof. Once the set of the significant interference sources is determined by each femtocell, then the resource slots for femtocell is determined by $\frac{M-m_j}{M}$ where $m = |S_j|$. On the one hand, no femtocell can increase its utility by decreasing its power. On the other hand, in the allocated resource blocks, no femtocell user is motivated to increase its transmission power because it would definitely violate the interference at the most vulnerable macro user in $U \setminus S_j$ and be shutdown without no gain.

References

1. Chandrasekhar V, Andrews JG (2009) Uplink capacity and interference avoidance for two-tier femtocell networks. IEEE Trans Wireless Comm 8(7):3498–3509
2. Small Cell Forum (2014) Document 049.03.01 Backhaul technologies for small cells. In: Release three: urban foundations. Small Cell Forum. Available via DIALOG. http://www.scf.io/en/documents/049_Backhaul_technologies_for_small_cells.php. Accessed 27 Feb 2014
3. Song W, Zhuang W (2012) Interworking of wireless LANs and cellular networks. Springer briefs in computer science. Springer, New York. ISBN: 978-1461443797
4. Pawelczak P, Venkatesha R, Xia L, Niemegeers IGMM (2005) Cognitive radio emergency networks: requirements and design. Proc IEEE DySPAN'05, Baltimore, pp 601–606
5. Kim H, Shin KG (2008) Efficient discovery of spectrum opportunities with MAC layer sensing in cognitive radio networks. IEEE Trans Mobile Comput 7(5):533–545
6. Cai LX, Shen X, Mark JW, Cai L, Xiao Y (2006) Voice capacity analysis of WLAN with unbalanced traffic. IEEE Trans Vehicle Technol 55(5):752–761
7. International Telecommunicaion Union (1996) General characteristics of international telephone connections and international telephone circuits one-way trans mission time. ITU, Geneva
8. Video Traces Research Group (2000) Video traces for network performance evaluation. In: Video trace libarary. Arizona State University, Tempe. http://trace.kom.aau.dk/TRACE/pics/FrameTrace/h263/indexc366.html. Accessed 10 Feb 2014
9. Andrews JG, Claussen H, Dohler M, Rangan S, Reed MC (2012) Femtocells: past, present, and future. IEEE J Sel Areas Comm. 30(4):497–508
10. Zhang J, Zhang Q (2009) Stackelberg game for utility-based cooperative cognitive radio networks. In: Proceedings of ACM MobiHoc'09, New Orleans
11. Taranto RD, Popovski P, Simeone O, Yomo H (2010) Efficient spectrum leasing via randomized silencing of secondary users. IEEE Trans Wireless Comm 9(12):3739–3749

Chapter 5
Conclusions and Future Directions

In this chapter, we summarize the main concepts and results presented in this Brief and highlight future research directions.

5.1 Conclusions

In this Brief, we aim at improving the performance of a heterogeneous cellular access infrastructure with small cell deployment. The main content of this Brief is summarized as follows:

- We have investigated the small cell techniques in cellular communications and previous works related to HetNet deployment and management. A framework of cognitive cellular networks has been presented to improve the HetNet performance by applying cognitive radio techniques. Several research problems have been discussed including backhaul design and management, interference management and intra-/inter layer cell coordination. Moreover, the motivations and challenges related to these research issues have been highlighted.
- We have addressed routing and resource allocation issues in wireless backhaul of HetNet, which have been formulated as a multi-hop dynamic spectrum access problem in tiered CRN. A spectrum-aware opportunistic cognitive routing (OCR) protocol has been presented to utilize time/location varying spectrum access opportunities for transmissions in wireless backhaul. Furthermore, we have presented a joint channel and relay selection algorithm, and simulation results have shown our protocol and algorithm outperform peer routing solutions with less computational and communication overhead and better delay performance.
- We have presented and analyzed a distributed QoS-aware cognitive (QC) MAC with service differentiation for small cells in HetNet, which supports heterogeneous multimedia applications. The channel access decision made by each single small cell considers the activities of both macrocell users and small cell users. Effective power allocation has been discussed based upon the design, which

Y. Liu and X. Shen, *Cognitive Resource Management for Heterogeneous Cellular Networks*, SpringerBriefs in Electrical and Computer Engineering, DOI 10.1007/978-3-319-06284-6__5, © The Author(s) 2014

applies game theory to analyze and regulate the nodal behavior in favor of fair spectrum sharing and entire network benefit. Simulation results have validated the effectiveness of power allocation in constrained backhaul capacity, and shown that the QC MAC provides satisfactory QoS performance for multimedia applications.

5.2   Future Research Directions

The flexible cell deployment and diverse traffic demand lead to many new challenging research issues under the scenario of HetNet. We close this chapter by presenting our future works as research directions in this field.

- The successive miniaturization of cellular access nodes accelerates the deployment of low power low cost small cells, which visions a future for more personalized cellular services deployed in the vicinity of subscribers. Therefore, small cells should function in a cognitive way to better adapt to the working environment. For example, on-site awareness is required at SBS, which optimizes the operation parameters by collecting and utilizing the knowledge of traffic pattern (e.g., type, load and variations) and ambient spectrum usage pattern (e.g., noise level, interference type and spatial distribution) [1]. Statistics characterizing local usage of a cell should be obtained and maintained during the cell operation so that proactive strategy can be applied to predict and adapt to user demand (e.g., content preloading [2], bandwidth preservation).
- The popularity of private cellular access nodes greatly challenges cellular operators in network planning and management. Given the large number of residing small cells and diverse coordination capabilities, the control method in cellular networks should transfer from delicate scheduling relying on real time global channel observation to effective admission and control based upon distributed management at individual access nodes. The network manager, on the other side, should define and monitor key system parameters related to network health in HetNet, e.g., the admitted connection count at macroscope and the link delay fluctuation at microscope, as in transportation networks [3]. Responding methods should then take actions upon the detection of problems in the network.
- Fast and robust backhaul in HetNet works as the aggregation point for small cells, which bridges the gap between RAN and core networks for integrity in cellular systems. Backhaul deployment and management are currently still in the infancy with many challenging open issues, such as the growing number of SBSs connected wirelessly [4], more energy awareness to embrace the "greener" communications [5], and heterogeneous approaches to the edge of core networks with various latency and capacity limits [6]. The most urgent problem is that backhaul is still lack of unified measurement to quantify the capability of different solutions (from wired to wireless, from fibre optical to DSL, from licensed to secondary usage). Therefore, new network model and analysis on

backhaul capacity should be provided as the foundation of the discussion on adaptive cell selection and cooperative multi-point transmission in tired HetNet. In addition, security and privacy issues related to privately deployed access nodes are also an open problem, e.g., identity theft by rogue base station attacks [7].

We hope you enjoy reading the Brief and find it useful.

References

1. Haykin S (2005) Cognitive radio: brain-empowered wireless communications. IEEE J Select Areas Commun 23(2):201–220
2. Lu R, Lin X, Shen X (2010) SPRING: a social-based privacy-preserving packet forwarding protocol for vehicular delay tolerant networks. In: Proceedings of IEEE INFOCOM'10
3. Knoop VL, Van Lint JWC, Hoogendoorn SP (2012) Routing strategies based on the macroscopic fundamental diagram. Transport Res Rec 2315:1–10
4. Love DJ (2012) Wireless backhaul (and access) at millimeter wave frequencies. In: CTW2012. http://www.ieee-ctw.org/2012/CTW_2012_web.pdf. Accessed 27 Feb 2014
5. Cai LX, Poor HV, Liu Y, Luan H, Shen X, Mark JW (2011) Dimensioning network deployment and resource management in green mesh networks. IEEE Wireless Commun Mag 18(5):58–65
6. Song W, Zhuang W (2009) Multi-service load sharing for resource management in the cellular/WLAN integrated network. IEEE Trans Wireless Commun 8(2):725–735
7. Perez D, Pico J (2011) A practical attack against GPRS/EDGE/UMTS/HSPA mobile data communications. In: Black Hat DC 2011. Taddong. Available via DIALOG. https://media.blackhat.com/bh-dc-11/Perez-Pico/BlackHat_DC_2011_Perez-Pico_Mobile_Attacks-wp.pdf. Accessed 27 Feb 2014